WORLD HEALTH ORGANIZATION

INTERNATIONAL AGENCY FOR RESEARCH ON CANCER

THE ROLE OF THE REGISTRY IN CANCER CONTROL

EDITORS

D.M. PARKIN   G. WAGNER   C. MUIR

IARC Scientific Publications No. 66

INTERNATIONAL AGENCY FOR RESEARCH ON CANCER
LYON
1985

The International Agency for Research on Cancer (IARC) was established in 1965 by the World Health Assembly, as an independently financed organization within the framework of the World Health Organization. The headquarters of the Agency are at Lyon, France.

The Agency conducts a programme of research concentrating particularly on the epidemiology of cancer and the study of potential carcinogens in the human environment. Its field studies are supplemented by biological and chemical research carried out in the Agency's laboratories in Lyon and, through collaborative research agreements, in national research institutions in many countries. The Agency also conducts a programme for the education and training of personnel for cancer research.

The publications of the Agency are intended to contribute to the dissemination of authoritative information on different aspects of cancer research.

Distributed for the International Agency for Research on Cancer
by Oxford University Press, Walton Street, Oxford OX2 6DP

London   New York   Toronto
Delhi   Bombay   Calcutta   Madras   Karachi
Kuala Lumpur   Singapore   Hong Kong   Tokyo
Nairobi   Dar es Salaam   Cape Town
Melbourne   Auckland

Oxford is a trade mark of Oxford University Press

Distributed in the United States
by Oxford University Press, New York

ISBN 0 19 723066 0
ISBN 92 832 1166 9 (Publisher)
© International Agency for Research on Cancer 1985

The authors alone are responsible for the views expressed in the signed articles in this publication. All rights reserved. No part of this publication may be reproduced, stored in a retrieval system, or transmitted, in any form or by any means, electronic, mechanical, photocopying, recording, or otherwise, without the prior permission of Oxford University Press

PRINTED IN SWITZERLAND

CONTENTS

Foreword v

Introduction vii

1. Cancer Registration: Historical Aspects
   G. Wagner 3
2. The Cancer Registry in Cancer Control: An Overview
   C.S. Muir, E. Démaret & P. Boyle 13
3. Planning and Evaluating Preventive Measures
   L. Teppo, M. Hakama, T. Hakulinen, E. Pukkala & E. Saxén 27
4. Evaluating and Planning Screening Programmes
   D.M. Parkin & N.E. Day 45
5. The Cancer Registry as a Tool for Detecting Industrial Risks
   O.M. Jensen 65
6. Planning Services for the Cancer Patient
   R.J. Wrighton 75
7. Survival Rate as an Index in Evaluating Cancer Control
   A. Hanai & I. Fujimoto 87
8. Cancer Care Programmes: The Swedish Experience
   T.R. Möller 109
9. Service Role of the Hospital Tumour Registry in the USA
   C. Zippin & M. Feingold 121
10. Second Cancers as a Result of Cancer Treatment
    N.E. Day & G. Engholm 133
11. The Role of Cancer Registration in Developing Countries
    C.L.M. Olweny 143

Index 153

FOREWORD

The cancer registry is a familiar resource to the epidemiologist, yielding information on the risks of cancer in different population groups, and on the changes that occur with time, from which etiological hypotheses can be developed. The registry provides an economical mechanism for following up industrial and other cohorts of individuals with specific exposures, and may be a useful source of subjects for case-control studies.

The information collected by registries can and should, however, be used in many other ways, notably in the planning and evaluation of cancer control programmes. Cancer control includes not only the search for epidemiological risk factors, since only by their identification can strategies of primary prevention be formulated, but also the provision of screening and early detection, therapy of established disease, and rehabilitation following treatment. Knowledge of the distribution and trends and the magnitude of the problem posed by different cancers is clearly essential in devising appropriate health care policies; monitoring the effectiveness of the measures taken requires similar information. This publication concentrates on the role that cancer registries play in these processes.

This series of papers derives from discussions that followed the 1983 Annual Scientific Meeting of the International Association of Cancer Registries, held in Heidelberg, Federal Republic of Germany. The theme of the meeting was 'The Benefits of Cancer Registration to the Cancer Patient and Society', and it was clear from many of the presentations that there was a wide variety of experience in the use of data derived from cancer registries in many aspects of cancer control, as well as in etiological research. Some of the contributions to this volume are revised versions of presentations given at the Heidelberg meeting, and a few have been specially commissioned to explore certain subject areas in greater depth. The result is a review which demonstrates the important part that the registry can play for clinicians, administrators and epidemiologists concerned with the fight against cancer.

L. Tomatis, M.D.
Director
IARC

INTRODUCTION

It is axiomatic that prevention of cancer is infinitely preferable to the treatment of established disease. In order to develop rational strategies of prevention, the risk factors or protective agents must be clearly identified.

Descriptive epidemiology suggests that perhaps 80% of cancer cases are environmentally determined (Higginson & Muir, 1979; Doll & Peto, 1981). The search for the causative agents involved relies upon analytical epidemiological studies of cancer risk in relation to individual exposure, which meet some or all of the criteria suggesting causality (MacMahon & Pugh, 1970). However, although we may be confident that most human cancer is environmentally determined, there remain large gaps in our knowledge of the precise components of the environment concerned; further, many of the dietary, social and cultural practices that have been etiologically incriminated often prove intractable to change. The scope for primary prevention may thus be somewhat less than it appears at first sight. The early detection of cancers and their effective treatment thus remain of great importance in the overall strategy of cancer control, and in cases for which there is no curative prospect, pain relief and the palliation of distressing symptoms are of immense importance.

This volume brings together several contributions which illustrate some of the ways in which data derived from cancer registries have been used in the planning and evaluation of services for cancer control. Since the use of cancer registries in epidemiological investigations of causes of cancer is fairly well known, its description is relatively brief, and emphasis is placed on their use in planning preventive programmes. In the latter context, the use of registry data in setting priorities, in examining the probable outcome of different interventions, in health education and in the evaluation of primary prevention is described. The special role of the registry in identifying occupational hazards and in the consequent development of preventive and legislative measures to ensure safer workplaces is described separately.

Planning of health care programmes for cancer requires knowledge of the size of the problem presented by different cancers, their distribution among different groups of the population, and past time trends and their probable evolution in the future. Incidence rates derived from cancer registration are the single most useful type of data for this purpose, although, in their absence, proxy measures such as mortality rates or the relative frequency of different cancers in hospital or pathology series may be used. When coming to decisions on the level of provision of services, information on the facilities that already exist for prevention, treatment and after-care (e.g., numbers of buildings, personnel, equipment) and their utilization is clearly pertinent, yet statistical data frequently play little part in the crucial decisions of how facilities or their utilization can be altered to achive the desired goals. However, in evaluating the effectiveness of different services for prevention, early detection and treatment of cancer, quantitative information on the improvements achieved is patently essential.

Most screening programmes for cancer detect early disease, with the objective of improving outcome of treatment and reducing the frequency of complications (including death) in the screened population. In contrast, screening for cervical cancer attempts to prevent the onset of invasive disease. The cancer registry can play an important part in evaluating the effectiveness of such screening programmes.

The registry has a significant role in several different areas of the care of cancer patients. At the level of individual hospitals, the registry can support clinicians in record keeping, case management and follow-up, and can provide statistical data on patients treated in hospital, which is of value in professional education and audit. Where cancer care programmes have been instituted, the cancer registry can make a particular contribution to some rather special features of patient management (World Health Organization, 1982).

Cancer registries have been most valuable in assessing the results and hazards of therapy. The most objective measure of the outcome of treatment is provided by the survival rate, and cancer registries are well placed to provide survival data for defined populations of individuals. The overall prognosis for different cancers cannot readily be judged from special series of cases entered into clinical trials; the results for whole populations are more meaningful. Cancer registries can be used to follow up groups of patients who have received treatments which are themselves potentially carcinogenic.

Cancer registries face many difficulties in fulfilling these diverse roles. These include the special problems of developing countries in maintaining adequate statistical systems to cope with the certain future increase in the relative and absolute size of their cancer problem. Even in those countries where registries have been established for many years, there may be a reluctance to provide adequate funding, so that coverage of a population is often limited (to certain towns or provinces), and the availability of information may be unnecessarily delayed due to lack of personnel to analyse the data amassed.

Finally, the operation of many cancer registries is threatened by increases in regulations and legislation designed to ensure protection of individuals and to prevent cross-linkage of different data files. Identification of individuals (by name, or unique number) is an essential part of the registration process, and linkage of records from different sources is necessary for estimation of survival and for follow-up studies. It is essential that the special needs of cancer registries be recognized when data protection codes are formulated. To quote Sir Richard Doll (1977), 'in my experience, most people understand that we cannot protect them against disease unless we are allowed the necessary tools'. This publication demonstrates just what a useful tool the cancer registry can be.

The Editors

REFERENCES

Doll, R. (1977) Strategy for detection of cancer hazards to man. *Nature,* **265,** 589–569

Doll, R. & Peto, R. (1981) *The Causes of Cancer,* Oxford, Oxford University Press

Higginson, J. & Muir, C.S. (1979) Environmental carcinogenesis: misconceptions and limitations to cancer control. *J. natl Cancer Inst.,* **63,** 1291–1298

MacMahon, B. & Pugh, T.F. (1970) *Epidemiology, Principles and Methods,* Boston, Little, Brown & Co., pp. 20–23

World Health Organization (1982) *Development of Cancer Centres and Community Cancer Control Programmes (EURO Reports and Studies No. 70),* Copenhagen

THE ROLE OF THE
REGISTRY
IN CANCER CONTROL

1. CANCER REGISTRATION: HISTORICAL ASPECTS

G. WAGNER

German Cancer Research Center, Heidelberg, Federal Republic of Germany

The prototype of the epidemiological cancer registry in its present form did not arise like a phoenix from the ashes, but has developed as a slow process, with many detours and blind alleys, from varying preliminary stages. Identifying the historical sources of the cancer registry is thus largely a matter of subjective discretion.

The oldest form of registration is the census; such counts were carried out in China as early as 500 BC, and, according to Logan and Lambert (1975), some sort of registration is supposed to have existed under Pharaoh Ramses II (*ca* 1250 BC). The Bible mentions several such censuses; the best known is that of Herod at the command of the Emperor Augustus at the time of the birth of Christ.

The biblical myth tells us that Jehovah's rage was vented against the census of King David:

'And David's heart smote him after that he had numbered the people. And David said unto the Lord: "I have sinned greatly in what I have done; but now, O Lord, put away, I beseech Thee, the iniquity of Thy servant; for I have done very foolishly." And when David rose up in the morning, the word of the Lord came unto the prophet Gad, David's seer, saying: "Go and speak unto David: Thus says the Lord: I lay upon thee three things; choose thee one of them that I may do it unto thee." So Gad came to David and told him, and said unto him: "Shall seven years of famine come unto thee in thy land? or wilt thou flee three months before thy foe while they pursue thee? or shall there be three days' pestilence in thy land? now advise thee, and consider what answer I shall return to Him that sent me." And David said unto Gad: "I am in a great strait; let us fall now into the hand of the Lord; for His mercies are great; and let me not fall into the hand of man."
So the Lord sent a pestilence upon Israel from the morning even to the time appointed; and there died of the people from Dan even to Beer-sheba seventy thousand men.'

(Samuel II, chapter 24)

The well-known German statistician Wagemann (1950) in his book *Narrenspiegel der Statistik (Fool's Mirror of Statistics)* expressed the opinion that in this citation from the Bible theological arguments are brought forward against a royal decree which was sure to be most unpopular with the Jewish people; for what else could King

David intend with the census than the recruitment of men capable of bearing arms or the introduction of new taxes? Perhaps the rather widespread antipathy against registration has one of its last roots in the fear of the exchequer (Wagner, 1957).

In the old city of Rome, newborn children had to be registered 30 days after birth; thus there already existed kinds of registrars who registered births and deaths. With the decline of the Roman Empire, these activities, too, were lost.

Only at the beginning of the sixteenth century did registration again begin, at first in the parishes of England. As of 1538, every clergyman of the 'Established Church of England' had to keep a book into which, in the presence of the churchwarden, all weddings, baptisms and funerals of the past week had to be entered. As early as 1532, 'Bills of Mortality' existed in London, which consisted of weekly reports on the deaths that had occurred, with particular consideration to deaths from epidemics. John Graunt (1620–1674) drew upon the bills for the years 1604–1661 in order to publish his famous book *Natural and Political Observations Mentioned in a Following Index and Made upon the Bills of Mortality*, in which he described, among other things, the excess of male births, the high infant mortality and the seasonal variation of mortality and, for the first time, constructed a life table. Because of his pioneer work in the application of probability theory to mortality data, Graunt is widely considered to be the founder of biostatistics (Lilienfeld & Lilienfeld, 1980), the founder of demography (Glass, 1963), or a forerunner of epidemiology (MacMahon & Pugh, 1970).

An early pioneer in Germany was Johann Peter Süssmilch (1707–1767), who, in his three-volume work *Die göttliche Ordnung in den Veränderungen des menschlichen Geschlechts... (The Divine Order in the Changes of the Human Species...)*, which appeared in 1741, set up statistical modes of population development whose laws he considered to be the expression of divine action. Süssmilch (1741)—an army chaplain and later a parson—mentioned in the preface to the first volume that he had studied the work of Graunt and wrote: 'Dem Herrn Graunt gebührt das erste und vornehmste Lob, als welcher in diesen neuen Wahrheiten das Eis gebrochen und welcher zuerst die bis zu seiner Zeit in London gedruckten Listen zu deren Entdeckung zu nutzen gesuchet.' (The first and most noble praise is due to Mr Graunt who broke the ice with these new truths and who was first to endeavour to use the lists printed in London in his time to discover these truths).

Going beyond Graunt, Süssmilch analysed numerous aspects of population development and its biological and economic causes, and formulated, in cooperation with the famous Swiss mathematician Leonhard Euler, his own methods of population statistics.

The collection and analysis of medical data was given a new impetus by the French Revolution. The key figure in the dissemination of epidemiological ideas was Pierre Charles-Alexandre Louis (1787–1872), the intellectual father of the so-called 'numerical method'; his contemporary, Jules Gavarret, author of the famous book *Principes généraux de statistique médicale*' (1840), should also be mentioned (Lilienfeld, 1979).

Two of Louis' pupils—William Augustus Guy (1810–1885) and William Farr (1807–1883)—greatly influenced the further development of epidemiology in England in the middle of the nineteenth century. Guy became one of the most important physicians of his time and a pioneer in the field of occupational epidemiology. Farr, after trying unsuccessfully to establish a private practice (Lilienfeld, 1979), was employed as 'Compiler of Abstracts' in the Registrar General's Office in 1839 and soon became the outstanding figure of this office, contributing to the development of epidemiological concepts. He coined, among others, the terms 'prevalence', 'standardized mortality rate', 'person-years', 'dose-response effect'. He was probably the first to suggest retrospective and prospective studies, introduced the first modern national population registration system and, together with d'Espine, founded the International Classification of Diseases (1853).

Contemporaries of Farr—his assistant John Simon (1816–1904), co-founder of the London Epidemiological Society (1850) and first 'Medical Officer of Health' in England (Hobson, 1975), as well as Lemuel Shattuck (1793–1859) in Boston, Massachusetts, the first promoter of a general population registration in the USA (Rice, 1981)—also deserve mention.

Until the beginning of this century, the epidemiology of cancer was based exclusively on mortality statistics. In England and Germany, in particular, critical voices were heard at the turn of the century with regard to the increase in the number of cancer deaths, demanding improved statistics on the spread of cancer in the population as a basis for research into its causes. In Hamburg, Dr Alexander Katz (1899) demanded a 'Sammelstatistik über Krebserkrankungen' (collective statistics on cancer). A year later, in Berlin, the 'Comité für Krebsforschung' (committee for cancer research) was constituted under the chairmanship of Professor Ernst von Leyden, which attempted to register all cancer patients in Germany then under medical treatment. After four preparatory meetings of the committee, two questionnaires were sent to every physician in the country *via* the Prussian Ministry of Culture, the first of which contained questions about the personality of the patient and his cancer disease, the second of which sought data on the patient's environment. This campaign was carried out as a 'key-day' investigation on 15 October 1900 (Comité für Krebsforschung, 1901).

The German approach was met with great interest and was repeated in various European countries. On the same day, an identical investigation was carried out in the Netherlands, using the German questionnaires as a model, and similar studies were carried out in 1902 in Spain (Leyden, 1904), in 1904 in Portugal and Hungary, in 1905/06 in Sweden and in 1908 in Denmark and Iceland (Fibiger & Trier, 1910) (Table 1).

In the report on the campaign in Germany, it was noted that 'a little more than half of the physicians addressed' had filled and returned the questionnaires (von Leyden *et al.*, 1902). In 165 areas with strikingly high numbers of cancer patients, a supplementary campaign was carried out in November 1902, which included questions particularly concerning living conditions, nutrition and family of the cancer patients. Two censuses, one in Württemberg (1931) and one in Baden (1930/34), which lasted six months and in which 97% of all physicians participated, were particularly successful. A five-year census of cancer patients was carried out in the cities of Göttingen, Halle, Kiel and Nuremberg from 1 October 1933 to 30 September 1938 (Lasch, 1940). Hospitals and practising physicians reported to statistical offices where the data were further processed; however, since the reports were anonymous, they could not be checked or corrected.

Table 1. Early censuses of cancer patients in Europe[a]

Year	Country
1900	Germany
1900	Holland
1902	Spain
1904	Portugal
1904	Hungary
1904/06	Germany (Baden)
1905/06	Sweden
1908	Denmark
1908	Iceland

[a] From Lasch (1940)

Considering the results of such surveys unsatisfactory, however, F.C. Wood (1930), director of the Institute of Cancer Research at Columbia University, New York City, demanded in 1930 that cancer be made a notifiable disease and that compulsory registration of all cancer cases be introduced.

The 'laufende fortgeschriebene individuelle Morbiditätsstatistik' (current extrapolated individual morbidity statistics) conceived in 1940 by C.H. Lasch, a radiologist from Rostock, constituted a methodological improvement in so far as the patient rather than the case constituted the unit of enumeration; reporting by name made it possible to eliminate multiple registration and to determine individual outcomes. This kind of morbidity statistics already approximates to the present form of cancer registration.

Continuous collection of cancer morbidity statistics began in Mecklenburg in 1937. All medical practitioners, hospitals and pathological institutes received registration cards or registration forms which had to be filled for cancer patients stating treatment and sent to the pertinent statistical office every two weeks. There, the reports were checked and entered into a card index of new patients. Missing reports were traced (eight days after the reporting date had elapsed, by daily reminders over the telephone!). This method seems to have worked fairly well, as indicated by the rate of coverage, which in 1937/38 was about 200 new patients per 100 000 inhabitants.

Following this favourable experience, analogous investigations were instituted in Saxony-Anhalt, in the Saarland and in Vienna in 1939; these soon had to be discontinued, however, because of political developments. Only in Mecklenburg and in Vienna did these investigations continue during the first years of the war.

In the USA, incidence data had been assembled by ad-hoc morbidity surveys on a national scale; the first such survey was conducted in 1937–1939, the second in 1947–1948 and the third in 1969–1971 (Haenszel, 1975). The sole purpose of these investigations was to obtain epidemiologically useful data for cancer control; the fate of the patients was of secondary interest.

Some of the oldest cancer registries in existence today started with the opposite approach: for them, follow-up and aftercare of the individual patient was of primary importance; the statistical material obtained was, in a way, a by-product. An example of such a registry is that of Hamburg, which is, I believe, the oldest registry in existence. Starting with the idea that cancer control involves not only medical and scientific but also public health and economic aspects, the municipal physician Professor Dr G.H. Sieveking, together with the director of the Hamburger Forschungsinstitut für Krebs und Tuberkulose (Hamburg Research Institute for Cancer and Tuberculosis) in Eppendorf, Professor Dr R. Bierich, founded an after-care organization for cancer patients. At the beginning, in 1927, this service worked on a private basis; from 1 January 1929 it obtained official status as the Nachgehender Krankenhilfsdienst (Follow-up Patient Care Service), a department of the Hamburg Public Health Office (Bierich, 1931; Keding, 1973). Three nurses were employed who visited hospitals and medical practitioners in Hamburg at regular intervals, recorded the names of new cancer patients and transferred relevant data to a central card index in the Health Office. In order to keep the card index up to date, it was compared weekly with official death certificates. Sieveking (1930, 1933, 1935, 1940) produced several reports on the Service, to which the reader is referred. The central card index and its statistical evaluation (carried out later in collaboration with the Statistical Office of the City of Hamburg) formed the basis of the Hamburg Cancer Registry.

The first modern registries for epidemiological purposes were set up shortly before and during the Second World War. In the USA, the first, and still in many respects exemplary, cancer registry was created in the state of Connecticut in 1936 (Connelly et al., 1968). The state of New York (without New York City) followed in 1940 (Petrakis, 1979). The first registry in Canada was opened up in the province of Saskatchewan in 1944; and in Europe registries were started in Denmark (1942), Belgium (1943) and the UK (south-western region, 1945). In New York State and in Saskatchewan, registration was compulsory, on a legal basis, from the very beginning; the other registries functioned on the basis of voluntary reporting on the part of hospitals and physicians.

Probably the most important impetus for the worldwide establishment of today's cancer registries came from a conference that took place in Copenhagen from 2 to 6 September 1946 upon the initiative of J. Clemmesen (Schinz, 1946). The twelve participants of the conference—leaders in the field of cancer control—emphasized the need for setting up improved and internationally comparable morbidity statistics, using uniform nomenclature and classification as well as an ideal standard population. To reach this goal, they recommended the worldwide establishment of epidemiological cancer registries.

The resolution accepted at the Copenhagen meeting reads as follows:

Whereas it is universally admitted:
1. That there are considerable variations in death rate from cancer of specific organs according to occupation and social status;
2. That there is variation in the incidence of cancer of the cervix and cancer of the corpus of the uterus in different races, and between parous and nulliparous women;
3. That there is a difference in the proportion between the number of deaths caused by cancer of the stomach and other sites in different countries and parts of countries;

4. That the number of deaths verified as carcinoma of the lung is increasing in many countries;
5. That the collection of accurate statistics of cancer incidence and mortality amongst different classes of people and in different towns and countries may lead to important indications for experimental studies;
6. That the information at present available is becoming increasingly inadequate owing to the growing numbers of patients successfully treated and, thus, not registered in the statistics of death,

We, the undersigned, meeting at Copenhagen, 6th September 1946, make the following suggestions:
1. That great benefit would follow the collection of data about cancer patients from as many different countries as possible;
2. That such data should be recorded on an agreed plan so as to be comparable;
3. That each nation should have a central registry to arrange for the recording and collection of such data;
4. That there should be an international body whose duty it should be to correlate the data and statistics obtained in each country.

If these suggestions are accepted, we further suggest:
1. That the international organisation should advise a terminology and methods of classification and tabulation which should be used by all cooperating countries;
2. That to obtain the data a registration of all cancer cases, whether compulsory or voluntary, would be necessary and a fee be paid to the doctor for such registration;
3. That each national registry should arrange to obtain information about each registered patient
   a) whether the patient was treated and, if so, by what method;
   b) follow-up notices from hospitals, doctors, etc.;
   c) by death certificate specifying occupation, the data of the first symptom, and the result of any autopsy;
4. That international comparisons of morbidity and mortality from cancer should be made in quinquennial groups of ages, or, if all ages are combined, an agreed standard population should be used.

'Signed: E.G.E. Berven (Stockholm); R.B. Engelstad (Oslo); I. Busk (Copenhagen); J. Clemmesen (Copenhagen); H. Karplus (Tel Aviv); F. Møller (Copenhagen); A.H.T. Robb-Smith (Oxford); D.W. Smithers (London); G.F. Stebbing (London); H.R. Schinz (Zurich); P. Stocks (London); W.F. Wassink (Amsterdam)'

The World Health Organization took up these recommendations, and an Expert Committee on Health Statistics founded a 'Subcommittee on the Registration of Cases of Cancer as well as their Statistical Presentation', which met for the first time in Paris in March 1950. At this and further meetings in Paris (1951) and Geneva

Table 2. Epidemiological cancer registries established before 1960 that are still operating

Country (region)	Year of establishment	Notification	Reference
Germany (Hamburg)	1929	Voluntary	Keding (1973)
USA (Connecticut)	1936	Voluntary	Griswold et al. (1955)
USA (New York)	1940	Compulsory	Burnett (1976)
Denmark	1942	Voluntary	Clemmesen (1965)
Belgium	1943	Voluntary	Jahn (1964)
Canada (Saskatchewan)	1944	Compulsory	Barclay (1976)
England and Wales	1945	Voluntary	Stocks (1959)
New Zealand	1948	Compulsory	Foster & Fraser (1982)
USSR	1948	Compulsory	Jahn (1964)
Yugoslavia (Slovenia)	1950	Compulsory	Ravnihar (1960)
Hungary	1952	Compulsory	Vikol (1966)
Norway	1952	Compulsory	Pedersen & Magnus (1959)
German Democratic Republic	1953	Compulsory	Wildner (1959)
Finland	1953	Compulsory	Saxén & Hakama (1965)
The Netherlands	1953	Voluntary	Jahn (1966)
Iceland	1954	Voluntary	Bjarnason & Tulinius (1983)
Sweden	1958	Compulsory	Ringertz (1971)
Japan (Miyagi)	1959	Voluntary	Segi (1966)
Israel	1960	Voluntary	Steinitz & Tzur (1965)
Spain (Zaragoza)	1960	Voluntary	Zubiri (1982)

(1957), recommendations were worked out for the establishment of epidemiological cancer registries (Stocks, 1959).

These beginnings have led to the establishment of nearly 100 epidemiological cancer registries throughout the world (Wagner & Ott, 1975). Table 2 gives the chronology of such cancer registries that were begun before 1960.

As a consequence of increasing specialization, a few epidemiological registries have developed lately which cover not all, but only very specific forms of cancer. They include:

– the registries of paediatric tumours in Houston, Mainz, Oxford and Sydney;
– the leukaemia registry in Perth;
– the bone tumour registry in Heidelberg;
– the registries of gynaecological tumours in Manchester and Louisville;
– the registries of gastrointestinal tumours in Dijon, Heidelberg and Prague;
– the retinoblastoma registry in Tokyo.

Despite the many efforts at international cooperation, however, it subsequently transpired that, for many reasons, the figures obtained in different countries could not be compared as well as had been hoped. It is at this point that the—at present —last chapter in the history of the cancer registry begins, namely the foundation of our International Association of Cancer Registries.

Within the framework of the Ninth International Cancer Congress held in Tokyo in 1966, about 60 interested scientists met at the invitation of Professor Segi for an informal meeting on cancer registration. In the final discussion, S. Cutler (National Cancer Institute, Bethesda, MD, USA) proposed the foundation of an international association of cancer registries whose task should be to coordinate and to supervise the content and methods of data collection on cancer patients. This suggestion was unanimously approved by all present. W. Haenszel was elected chairman of a preparatory organization committee; he conducted further negotiations, which, in May 1968, led to the official foundation of the association in Lausanne. The first president of the IACR was E. Pedersen (Oslo), the general secretary S. Cutler and the first honorary member Dr M.H. Griswold, retired director of the Connecticut Cancer Registry. The current president of the Association is Dr P. Correa (New Orleans, LA, USA), and the secretary is Dr C.S. Muir (Lyon, France).

This review has attempted to trace the growth of cancer registration from its earliest roots. Cancer registries are not a modern playground for crazy epidemiologists but one of the most important prerequisites of future progress in the fields of cancer control and cancer etiology, which have long since proven their worth.

REFERENCES

Barclay, T.H.C. (1976) *Canada, Saskatchewan*. In: Waterhouse, J., Muir, C., Correa, P. & Powell, J., eds, *Cancer Incidence in Five Continents Vol. III (IARC Scientific Publications No. 15)*, Lyon, International Agency for Research on Cancer, p. 160

Bierich, R. (1931) *Die Krebsbekämpfung in Hamburg*. In: Grüneisen, F., ed., *Jahrbuch des Reichsausschusses für Krebsbekämpfung*, Leipzig, J.A. Barth, pp. 47–48

Bjarnason, O. & Tulinius, H. (1983) Cancer registration in Iceland 1955–1974. *Acta pathol. microbiol. immunol. scand.*, Section A, Suppl. 281

Burnett, W.S. (1976) *USA, New York State (less New York City)*. In: Waterhouse, J., Muir, C., Correa, P. & Powell, J., eds, *Cancer Incidence in Five Continents Vol. III (IARC Scientific Publications No. 15)*, Lyon, International Agency for Research on Cancer, p. 224

Clemmesen, J. (1965) *Statistical Studies in Malignant Neoplasms*, Vol. 1, Copenhagen, Munksgaard

Comité für Krebsforschung (1901) *Verhandlungen*, Supplement to *Deutsche Medizinische Wochenschrift*

Connelly, R.R., Campbell, P.C. & Eisenberg, H. (1968) *Central Registry of Cancer Cases in Connecticut (Publ. Health Rep. 83, No. 5)*, Washington DC, US Public Health Service, pp. 386–390

Fibiger, J. & Trier, S. (1910) Bericht über die Zählung der am 1. April 1908 in Dänemark in ärztl. Behandlung gewesenen Krebskranken. *Z. Krebsforsch.*, **9**, 275–336

Foster, F.H. & Fraser, J. (1982) *New Zealand*. In: Waterhouse, J., Muir, C., Shanmugaratnam, K. & Powell, J., eds, *Cancer Incidence in Five Continents Vol. IV (IARC Scientific Publications No. 42)*, Lyon, International Agency for Research on Cancer, p. 608

Glass, D.V. (1963) John Graunt and his natural and political observations. *Proc. R. Soc. Biol.*, **159**, 2–37

Griswold, M.H., Wilder, C.S., Cutler, S.J. & Pollack, E.S. (1955) *Cancer in Connecticut 1935–1951*, Hartford, CN, Connecticut State Department of Health

Haenszel, W. (1975) The United States network of cancer registries. In: Grundmann, E. & Pedersen, E., eds, *Cancer Registry*, Berlin, Springer

Hobson, W. (1975) *The Theory and Practice of Public Health*, 4th ed., London, Oxford University Press

Jahn, E. (1964) Der Gedanke des 'Krebsregisters' und die Möglichkeiten seiner Verwirklichung in der BRD. *Bundesgesblatt*, **25**, 385–394

Jahn, E. (1966) Krebsregister. In: Wagner, G., ed., *Krebs – Dokumentation und Statistik maligner Tumoren*, Stuttgart, F.K. Schattauer, pp. 75–86

Katz, A. (1899) Die Notwendigkeit einer Sammelstatistik über Krebserkrankungen. *Dtsch. med. Wschr.*, **25**, 260–261, 277

Keding, G. (1973) Annotation zur Krebsepidemiologie. *Hambg. Ärztebl.*, **27** (No. 8)

Lasch, C.H. (1940) Krebskrankenstatistik. Beginn und Aussicht. *Z. Krebsforsch.*, **50**, 245–298

von Leyden, E., Kirchner, Wutzdorf, von Hansemann, & Meyer, G., eds (1902) *Bericht über die vom Komitee für Krebsforschung am 15. Oktober 1900 erhobene Sammelforschung*, Jena, Gustav Fischer

Leyden, H. (1904) Bericht über die am 1. September 1902 in Spanien veranstaltete Krebssammelforschung. *Z. Krebsforsch.*, **1**, 41–72

Lilienfeld, A.M. & Lilienfeld, D.E. (1980) *Foundations of Epidemiology*, 2nd ed., New York, Oxford University Press

Lilienfeld, D.E. (1979) The greening of epidemiology. Sanitary Physicians and the London Epidemiological Society (1830–1870). *Bull. Hist. Med.*, **52**, 503–528

Logan, W.P.D. & Lambert, P.M. (1975) Vital statistics. In: Hobson, W., ed., *The Theory and Practice of Public Health*, 4th ed., London, Oxford University Press

MacMahon, B. & Pugh, T.F. (1970) *Epidemiology. Principles and Methods*, Boston, Little, Brown & Co.

Pedersen, E. & Magnus, K. (1959) *The Incidence of Cancer in Norway 1953–1954*, Oslo, Norwegian Cancer Society

Petrakis, N.L. (1979) Historic milestones in cancer epidemiology. *Sem. Oncol.*, **6**, 433–444

Ravnihar, B.V. (1960) Eight years of cancer registration in Slovenia (Yugoslavia). *Acta unio int. cancrum*, **16**, 1578–1583

Rice, D.P. (1981) Health statistics: Past and present. *New Engl. J. Med.*, **305**, 219–220

Ringertz, N., ed. (1971) Cancer incidence in Finland, Iceland, Norway and Sweden. *Acta pathol. microbiol. scand.*, Section A, Suppl. 224

Saxén, E. & Hakama, M. (1965) *Cancer registration in Finland with a note on the 'dangers' of early diagnosis*. In: Proceedings of the European Extra-European Seminar on Cancer Prophylaxis and Prevention, Rome, September 17–19 1965, Vol. 1, Rome, Centro Soc. Stud. Precanc., pp. 145–151

Segi, M. (1966) Cancer registration in Miyagi Prefecture. Paper presented at the Special Meeting on Cancer Registries at the Ninth International Cancer Congress, Tokyo

Sieveking, G.H. (1930) Das Krebsproblem in der öffentlichen Gesundheitsfürsorge. *Z. Ges. verwalt. Ges. fürsorge*, **1**, 23–30

Sieveking, G.H. (1933) Die Hamburger Krebskrankenfürsorge 1927–1932. *Z. Ges. verwalt. Ges. fürsorge, 4,* 241–247

Sieveking, G.H. (1935) Die Hamburger Krebskrankenfürsorge im Vergleich mit gleichartigen in- und ausländischen Einrichtungen. *Bull. Schweiz. Ver. Krebsbekämpf., 2,* 115–123

Sieveking, G.H. (1940) Hamburgs Krebskrankenfürsorge 1927–1939. *Mschr. Krebsbekämpf., 8,* 49–52

Schinz, H.R. (1946) Kleine internationale Krebskonferenz vom 2.–6. Sept. 1946 in Kopenhagen. *Schweiz. med. Wschr., 76,* 1194

Steinitz, R. & Tzur, B. (1965) *Israel Cancer Registry. Its Data and Data Processing,* Jerusalem, Ministry of Health and Israel Cancer Association

Stocks, P. (1959) Cancer registration and studies of incidence by surveys. *Bull. World Health Organ., 20,* 697–715

Süssmilch, J.P. (1741) *Die göttliche Ordnung in den Veränderungen des menschlichen Geschlechts,* Berlin, J.C. Spencer (Facsimile Reprint, Kulterbuch-Verlag, Berlin, 1977)

Vikol, J. (1966) *Twenty-five years of cancer control in Hungary.* In: Vikol, J. & Sellei, C., eds, *Twenty-five Years in the Fight Against Cancer,* Budapest, State Oncological Institute

Wagemann, E. (1950) *Narrenspiegel der Statistik,* 3rd ed., Munich, Lehnen

Wagner, G. (1957) Bedeutung, Gefahren und Grenzen der Statistik in der Medizin. *Dtsch. med. Wschr., 82,* 1427–1432, 1484–1491

Wagner, G. & Ott, G. (1975) *Krebsregister.* In: Koller, S. & Wagner, G., eds, *Handbuch der medizinischen Dokumentation und Datenverarbeitung,* Stuttgart, Schattauer, pp. 1141–1155

Wildner, G.P. (1959) Aufbau und Organisation der Erfassung der Krebskranken in der DDR. *Dtsch. Ges. Wesen, 14,* 26–49

Wood, F.C. (1930) Need for cancer morbidity statistics. *Am. J. publ. Health, 20,* 11–20

Zubiri, A. (1982) *Spain, Zaragoza.* In: Waterhouse, J., Muir, C., Shanmugaratnam, K. & Powell, J., eds, *Cancer Incidence in Five Continents Vol. IV (IARC Scientific Publications No. 42),* Lyon, International Agency for Research on Cancer, p. 530

2. THE CANCER REGISTRY IN CANCER CONTROL: AN OVERVIEW

C.S. MUIR & E. DÉMARET

International Agency for Research on Cancer, Lyon, France

P. BOYLE

Department of Epidemiology and Biostatistics, Harvard School of Public Health, Boston, MA, USA

INTRODUCTION

Many roads lead to the control of cancer—among those walking them are the clinican treating the patient with established cancer, the cancer registry director assessing the burden of cancer in a population, the epidemiologist seeking to discover the causes, the legislator and the public health official deciding on and implementing preventive measures, and laboratory workers trying to uncover the mechanisms whereby the normal cell becomes malignant.

The purpose of this chapter is to show that the cancer registry is an essential part of any *rational* programme of cancer control, benefiting both the individual and the society in which he lives.

A cancer registry can be defined as an organization for the collection, storage, analysis and interpretation of data on persons with cancer. A hospital-based registry undertakes these tasks within the confines of a hospital or, increasingly, a group of hospitals. A population-based registry (discussed below) is concerned with all newly-diagnosed cases of cancer occurring in a population of well-defined composition and size.

THE USES OF THE CANCER REGISTRY

In the control of cancer, registries are rarely in the forefront, their tasks being rather in the nature of intelligence gathering about the current cancer burden in a community, providing the data needed to uncover the causes of cancer in humans and for evaluation of the effects of steps taken to control the disease.

Cancer control activities include:

(1) continued assessment of the levels of cancer in the population;

(2) provision of the personnel, hospital and other facilities and equipment needed for the diagnosis, treatment and rehabilitation of the cancer patient;
(3) evaluation of the effect of early diagnosis and of treatment;
(4) identification by epidemiological and laboratory studies of the initiating and promoting agents that cause cancer; and
(5) evaluation of the effect of removing initiators and promoters from the environment, or of enhancing resistance to them through immunization (as is postulated for primary liver cancer) or by use of micronutrients (retinol).

The role of the cancer registry in these facets of cancer control is examined below in a somewhat different order, the assistance given to the cancer patient being considered first.

Service to the cancer patient

No matter what legislation may be in force (Muir & Démaret, 1982), cancer registries ultimately depend on the willingness of individual members of the medical profession to report directly or indirectly newly diagnosed cases. As collaboration is always a two-way process, the registry has to provide a service in return. The information most appreciated is that of use to the physician in his daily work of patient care. This may include reminders that a given patient should be seen again on an anniversary date, that is, assistance with follow-up. Such reminders are increasingly produced automatically by computers and can be programmed to take account of the periodicity of recall considered appropriate by a given clinician. Such reminders may, on the request of the treating physician, be sent directly to the patient concerned but would give no indication that they originated in the cancer registry. The provision of annual listings of patients seen by an individual clinician or service is always welcome, as is help in preparing case series for publication. By centralizing records, cancer registries can assist physicians trying to determine previous treatment: patients do not always know their own history. The South Thames Cancer Registry (UK) automatically receives copies of death certificates on which there is a mention of cancer for deaths occurring in the registration area and sends an abstract to the respective hospital and treating physician so that records can be completed and, further, ensure that follow-up letters are not sent to deceased patients.

Survival and evaluation of treatment

The only way of knowing whether treatment A is better than treatment B is to mount a controlled clinical trial. Patients entered into such a trial are usually carefully chosen and are not likely to be representative of all patients with a particular type of cancer; hence, their survival rates give little indication of the fate of all patients. By matching all notifications sent to the registry with death certificates for the same area, the registry is able to assess the survival of *all* persons with cancer in the registration area, including those who for some reason are not treated at all (see Hanai and Fujimoto, this volume). Such figures, although usually depressing, do indicate how the average patient fares and serve to place in perspective extravagant claims about the success of therapy. While survival rates for choriocarcinoma, acute lymphatic leukaemia in children, Hodgkin's disease and cancer of the testis have improved substantially in the past five years, rates for common cancers such as those of lung, stomach and colon have generally not.

Some registries are willing to provide an analysis of the survival rates of patients seen in a given service, or by an individual clinician, so that these may be compared with rates for the registration area as a whole. Such information would not usually be published, and, if published as part of any analysis for the registration area, neither clinician nor service would be identified.

While comparison by the registry of the results of various treatment centres is possible, this requires great caution. For example, patients with advanced disease may be sent to the 'best' centre, with the consequence that this centre may have poorer results in terms of survival than those less well equipped and staffed. However, glaring differences between centres may point to a need for remedial action.

Several registries (e.g., that of Birmingham and West Midlands, UK) record the treatments given in some detail. Such information can be used to assess current treatment practice and forms a valuable point of departure when assessing the long-term consequences of chemotherapy and radiotherapy in terms of second neoplasms. To obtain sufficient numbers, it is advantageous to pool results from many registries, as exemplified by the recent publication on the risk of a second cancer subsequent to radiation treatment for cervical cancer (Day & Boice, 1984).

If appropriate information is collected and collated in a uniform manner (e.g., TNM, FIGO), it may be possible to assess survival for cancers of different size and degree of spread. However, problems in ensuring comparability usually result in presentation of survival as localized, regional spread and distant spread, as, for example, in Norway (Cancer Registry of Norway, 1980). Although estimation of survival by occupation or some other index of socio-economic level is possible, such analyses are rarely undertaken, as the validity of the information on occupation is frequently in doubt. The SEER (Surveillance, Epidemiology, and End Results) registries supported and coordinated by the National Cancer Institute in the USA record the census tract of residence, as this gives an indication of socio-economic level.

Provision of services

The information routinely collected by cancer registries permits assessment of the size of the problem and, hence, needs for hospital beds, staff and equipment. If data on time trends are available, then future needs can often be forecast. Rises in the incidences of lung cancer and breast cancer will clearly indicate the need for surgeons of differing training. Increases in cancers that are treated by radiation therapy or by chemotherapy will again indicate that different facilities be provided. It may be possible to detect stages at which there is undue delay before treatment starts, but such analyses require meticulous recording of date of first consultation, date of diagnosis, etc., if they are to have any value; an ad-hoc study of limited duration may be more useful. Monitoring of patterns of patient flow and referral may show when and where it would be cost-effective to open new treatment centres.

Evaluation of screening

Much money and effort has been expended worldwide on screening programmes for cervical (and more recently breast and large-bowel) cancer. Although the underlying reasoning is intuitively appealing—namely, the earlier the cancer is detected the greater the chance of cure—very few screening programmes have been put to the only valid test of efficacy, which is a lower death rate from cancer in the screened than among the nonscreened (Miller, 1978). The cancer registry, by vocation, is in an ideal position to assist in this assessment—if cases detected by the screening programmes are conscientiously notified to the registry. Results from the Danish, Finnish and Icelandic cancer registries have shown what can be done in this area (Hakama, 1982). Implicit in such evaluations is that the cancer registry have access by name to all death certificates and to the cause of death. Given that breast cancer screening, for example, is based on the use of radiation, which may by itself increase the risk of breast cancer, the need to know the fate of those screened is obvious.

Epidemiology: the size of the problem

One of the most important and probably the best known of the contributions that the cancer registry can make is to provide 'current' data on the incidence of cancer, by age, sex, place of birth, occupation, etc. (Information for a given year is usually available some two years later.) The computation of rates depends on the availability of relevant denominators. In countries where the means to take account of movements of populations between censuses are lacking, the population at risk in registration areas that are subnational in extent may become increasingly inaccurate towards the end of the intercensal period, and rates may have to be computed again when a new census becomes available.

Depending on the size of the area covered, regional or urban/rural differences may be demonstrable, risk being usually greater in urban areas (Roginski, 1982). These may be correlated with various features of the environment, resulting in hypotheses about etiology that can be tested by other means (Teppo *et al.*, 1980).

As the diagnosis of most cancers is based on the microscopic examination of a portion of the tumour, differences in histological type can also be examined, and these may be linkable to differences in etiology (Kreyberg, 1952; Muñoz *et al.*, 1968). In international studies, national differences in nomenclature and interpretation must be excluded before the conclusion is reached that any differences observed are real [for example, differences in the frequency of histological types of testicular cancer in the Birmingham (UK) and Connecticut (USA) cancer registries (Doll *et al.*, 1970)].

Once a registry has been established for some time, examination of time trends, particularly when done for birth cohorts (Laara, 1982), can provide unequalled insight into the likely size of the burden of a given site of cancer for many years to come. Thus, the decline in incidence of stomach cancer in Japan as a whole could be predicted some 20 years before it took place, since it was preceded by a fall in incidence in younger age groups (Muir *et al.*, 1981). Occasionally, cancer registries can detect or confirm the existence of an 'epidemic' of cancers, such as occurred for cancer of the corpus uteri in California following widespread use of oestrogens by menopausal women (Henderson *et al.*, 1983).

Epidemiology: the causes of cancer

If current ideas about the relative importance of the various causes of cancers (Wynder & Gori, 1977; Higginson & Muir, 1979; Doll & Peto, 1981) are correct, the 'easy' causes, such as tobacco smoking, excess alcohol consumption, radiation and a certain number of industrial exposures, have been discovered, and a much more difficult era requiring detailed investigation of the effects of lifestyle and dietary factors is beginning. The combined approaches of case-control and cohort (follow-up) studies will be needed.

Cross-sectional studies compare the incidence (or mortality) rate of cancer and the level of exposure to possible risk factors in different population groups. The demonstration of a correlation at the group level may reflect a causative link— examples are the relationship between fat consumption and cancers of the breast and colon (Armstrong & Doll, 1975), and that between aflatoxin exposure and hepatocellular carcinoma (Linsell, 1979). Such investigations may be useful for building hypotheses, but any correlation observed may reflect the effect of a linked exposure: e.g., heavy alcohol drinkers are often heavy smokers. Further, there is no guarantee that those members of a population who eat, say, larger amounts of fat are those who develop cancer. Hence, in order to study cancer risk in relation to different environmental exposures at the level of the individual, both case-control and cohort studies are necessary.

In the case-control approach, cancer patients and controls (persons without the cancer under study) are questioned about past exposures and the replies compared. Thus, a greater proportion of lung cancer patients will be found to be smokers than controls. In theory, the cancer registry provides an ideal source of subjects for case-control studies. Cases are representative of all those in the population, unbiassed by the selective factors that bring people to a single institution. However, in population-based registries, it is not usually possible to identify cases until several weeks or months after the time of diagnosis: for some cancers this may result in the introduction into the case series of bias due to factors that influence survival. Thus, cases derived from more limited sources are often used; however, the investigator can compare characteristics such as age, sex and residence of those interviewed with all those reported to the registry to see if the cases in the study were representative.

Although case-control studies are a powerful and economical way of testing hypotheses (Cole, 1979), many questions are better answered by the cohort approach. To give but one example, it is unlikely that an increased risk due to an industrial exposure could be identified by a case-control study, as the chance that cases and controls who work in that occupation are included in such a study would be very small—unless the investigation was carried out in a region where a large proportion of the workforce was employed in that industry. In a cohort study, a large number of persons without cancer are asked questions concerning their exposure to suspected risk factors, and these individuals are then followed to see which of them develop cancer. Thus, again, one would find that many more smokers develop lung cancer than nonsmokers. The great advantage of cohort studies is that replies are not distorted by the presence of disease, and it is possible to obtain current information, such as for diet, for which recollection tends to be poor. This method of investigation is frequently used to study industrial cancer risk.

Once the data have been collected, the epidemiologist must wait until sufficient cases of the diseases of interest appear in those participating in the study. Because this period of observation is usually long, there are many advantages to using existing routine statistical records systems to ascertain outcome, rather than to institute ad-hoc systems of follow-up. Offices of vital statistics record information on all deaths, including the cause of death, and have been widely used for this purpose—in the USA, for example, the National Death Index provides the computerized link between subjects in prospective studies and a file listing all decedents since 1979 (National Center for Health Statistics, 1981). However, death is now often postponed for several years by treatment; for many cancers a 'cure' is obtained, and death eventually occurs from causes other than the cancer in question. Incidence of cancer thus provides a much more satisfactory endpoint for cohort studies. When registries cover large areas they can be used to identify cases of cancer occurring in a defined cohort. For instance, the National Health Service Register in the UK has been used to 'flag' cohort members, and regional cancer registries also notify the Register of the occurrence of cancer in an individual, so that a linkage can be established (Office of Population Censuses and Surveys, 1981). Although linkage is manual, it is remarkably effective. The use of routine statistical sources for follow-up requires a system for matching cohort members with the files of registrations or deaths—and this involves the recording of names, or unique identifying numbers (see 'confidentiality' below).

Most cohort studies are designed to examine risk in an industry or group, either nationally or in several countries (Saracci *et al.*, 1984). When circumstances permit, the monitoring of risk for a sample, or for all, of the population defined at a census is increasingly practised (e.g., in England and Wales, Sweden).

If, as part of a cohort study, blood, urine or faeces are collected for storage at the same time as the questions are asked, such samples from those who develop cancer and from a small number of controls can be removed from storage and the levels of various chemicals measured. This, again, is a very economical method of answering questions, particularly for diet-linked cancers, if storage costs are reasonable and the chemicals of interest remain stable.

It cannot be emphasized too strongly that if the sole function of cancer registries were their use for cohort studies, their existence would be more than justified.

Intervention

The final proof of causation comes when removal of a suspected cause is followed by a fall in the cancer in question. For some rapidly fatal cancers, it may be sufficient to follow death rates, but for most, incidence figures from cancer registries are needed. When oestrogens were no longer widely prescribed in California, endometrial cancer incidence levels fell within six months (Henderson *et al.*, 1983), suggesting that these substances act as promoters. However, it usually requires years before the effects of intervention become apparent and quantifiable (Doll & Peto, 1976).

THE ADVANTAGES OF CANCER REGISTRATION

From time to time it is suggested that cancer registries are unnecessary and that most of their functions could be carried out by vital statistics offices using death certificates or hospital discharge summaries or a combination of these two sources. As such statements are made frequently, they are worthwhile examining in some detail.

Completeness

Since cancer is recorded on death certificates only when it is ascribed as a cause of death, and since survival varies widely for cancers at different sites, being nearly 100% for lip cancer and less than 10% for lung cancer, mortality rates cannot give a true picture of the burden of the disease. While weights reflecting survival can be applied to mortality data, this method of obtaining an approximation of incidence fails when there are changes in survival—as have recently occurred for cancer of the testis and Hodgkin's disease (De Vita *et al.*, 1980). Moreover, it takes some time before an improvement in survival can be quantified and a correction applied.

Cancer mortality data are generally published as part of a large, all-embracing vital statistics scheme in which a single disease or condition is selected as the underlying cause of death. While reasonably objective criteria (pathology, cytology, etc.) exist to define cancer, the attribution of 'cause' is frequently a more subjective phenomenon. Persons with cancer who die of a totally unrelated event are not included in mortality statistics: e.g., someone with breast cancer, with no evidence of disease three years after diagnosis and treatment, who dies of a stroke would not be counted.

Ascribing 'cause' gives rise to further problems. In a study in which 1246 US death certificates that mentioned cancer were sent to seven vital statistics offices to investigate the application of international rules for coding cause of death as they relate to cancer (Percy & Dolman, 1978), surprising differences were observed for a number of cancer sites. For example, for breast cancer there was remarkable variation between the Federal Republic of Germany (65) and France (95); for prostate cancer between England (39) and Norway (58); and for colon cancer between England (91) and Norway (116). Such inconsistencies are likely to be very much reduced when cancer notifications are coded in a registry.

Hospital discharge summaries or notifications would thus appear to be a more appropriate substitute. However, such summaries may not give the name, rarely distinguish between first and repeat admissions, and are frequently not linked from one year to another or between institutions. In other words, it may not be possible to distinguish incident and prevalent cases. Furthermore, the system may be biassed towards procedures (e.g., gastrectomy) rather than towards disease (e.g., gastric cancer), and coding at the hospital level may be of poorer quality than that at a cancer registry.

In summary, therefore, mortality data give a biassed underestimate of incidence, and hospital discharge data a biassed overestimate.

Accuracy

There have been several well known studies on the accuracy of statements of cause of death on death certificates (Heasman & Lipworth, 1966; Puffer & Wynne-Griffith, 1967), which show that diagnoses for individuals are often wrong. Thus, in cohort studies, in which the occurrence of cancer in an individual is important, the diagnosis recorded on a death certificate will clearly be less satisfactory than data on incidence from a cancer registry.

However, when the errors are corrected, the revised totals for a given site are often fairly close to the original numbers. Hence, for descriptive purposes mortality data may be adequate.

Even when the diagnosis of cancer has been made correctly in life, the death certificate may record a less precise or incorrect primary site. Percy *et al.* (1981) examined death certificates bearing mention of cancer of persons known to have cancer from the third National Cancer Survey (1968–1971) in the USA. While sites such as lung and pancreas were accurately reflected on the death certificates, cancers of larynx, bone and colon were overreported and those of cervix uteri, endometrium and rectum underreported.

Cancer registries are normally in constant dialogue with data sources and can often obtain more information about an individual. Vital statistics offices, covering all causes of death, are less likely to enquire. Thus, in Scotland, while nearly 12% of all deaths from colorectal cancer are coded to ICD-8 rubric 153.8 [cancer of large intestine (including colon)—part unspecified], less than 4% of incident cases are so coded.

Other problems of coding have been mentioned above. Coders of hospital discharge slips, who lack knowledge of the biology of the disease, frequently have difficulty in correctly coding statements mentioning metastases or multiple tumours.

Other considerations

Information regarding second primary tumours cannot be obtained from vital statistics offices. When an individual apparently cured of testicular cancer dies of iatrogenic leukaemia, no mention will be made of testicular cancer as the underlying cause of death (because in these circumstances it is not), since vital statistics schemes obey the dictum: one body, one cause. Only by coding the second part of the death certificate would the link be made following treatment. Cancer registration schemes, on the other hand, have been used very effectively to investigate the risk of second malignancy following treatment (see Day & Engholm, this volume).

Histological appearance is one of the features of interest to be registered for a cancer. Histology may have an influence on prognosis, e.g., basal or squamous-cell carcinoma of skin, and may reflect different etiology, e.g., angiosarcoma and hepatocellular carcinoma of the liver. Cancer registries record and code such information. Despite pleas from Dr J. Clemmesen of the Danish Cancer Registry that such information be required on death certificates, and although a code for the histology of tumours is given in the 9th Revision of the ICD, this step has not been taken by vital statistics offices.

Analytical epidemiology

Vital statistics offices have clearly no role in case-control studies, which require that the questions needed to test the hypothesis be posed to the cancer patient himself. Although information on certain items such as smoking has been obtained by proxy from relatives of deceased persons, the reliability of the responses is likely to be considerably less. As noted above, death certificates have been successfully used in cohort studies designed to demonstrate differences in risk between occupational groups, in which the end point is death from cancer. However, death certificates are much less satisfactory than cancer registry records in that only fatal cancers can be counted, and, further, as has been repeatedly shown, the quality of the information is poorer.

In summary, the wider applicability and accuracy of cancer registry data are preferable to those derived from death certificates. It is not a question of either/or, but rather of the use of both to derive the greatest knowledge about the disease and its causes.

CONFIDENTIALITY

In many countries today, there is a widespread desire to have an environment—notably the workplace, air, food and water—free from cancer risk. Indeed, political parties have emerged for which such matters are high on their list of priorities. At the same time there is a general wish that medical and other records not bear personal identification, so that they are entirely confidential and cannot be linked to other information about the same individual.

This desire has resulted in the creation of laws, regulations and data protection boards to protect the privacy of the individual. Recommendations on the confidentiality of individual data have also been established by supranational bodies like, for example, the Council of Europe (1981) and the Commission of the European Communities (1984). Confidentiality is also a major concern of the cancer registry, since, although the importance of cancer registration is recognized in most countries, the legal restrictions surrounding collection of information on individuals tend to make cancer registration and research based on registry and other medical information ever more difficult.

The incontrovertible fact is that linkage of individual records is essential if cancer is to be measured accurately, the causes ascertained at minimum expense and the effect of control measures assessed.

This is an issue that has to be faced squarely: one cannot have both absolute secrecy and rational effective cancer control. Some practical examples are given below.

If a cancer registration scheme is working properly, notifications will be received from several sources. Indeed, in Australia, the law requires this. To make sure that the same individual is counted once only, it must be possible to identify duplication—name, date of birth, sex and place of birth are essential for this pupose. A unique personal number is a useful adjunct, but such numbers can be forgotten or mixed up.

Cancer registries normally wish to find out if 'their' patients have died, in order to calculate survival. To match a name on a death certificate with those on the registry files is the most reliable and cheapest way of doing so. One of the methods used to assess completeness of registration is to ascertain the persons certified as having died from cancer for whom the registry had no record. Again, access by name to the death certificate is essential—yet in many countries this is forbidden by law.

In order to find out if cohort members have developed cancer (or some other disease), checking their names against cancer registry files and/or death certificates is again efficient and inexpensive. The alternative is to follow up each individual by letter, telephone, and, in some countries, by less creditable means such as detective agencies, driving licence records and credit bureaux, etc.—all vastly more expensive and, incidentally, involving the divulgation of information to third parties not bound by professional secrecy.

A shared confidence

Medical confidentiality is a shared confidence: treating physicians, nurses, radiologists, pathologists, laboratory technicians, ward maids, all know *who* has cancer. Why not the cancer registry?

The cancer registry may be asked to supply names of people with a given cancer so that they can be included in a case-control study. Names would not be divulged unless the attending physician gave his consent for each of his patients. Regrettably, some members of the medical profession seem to object and, indeed, when asked for permission to approach their patients for interview, fail to reply, thus in effect blocking the study. If permission is given, patient refusal is rare. Indeed, in our experience the cancer patient is pleased to be included in such studies, since he considers that they indicate an interest in his disease. (It is not necessary to mention the word cancer.) The control, if in hospital, has much the same motivation. Some 70% of randomly selected population controls are usually willing to answer questions.

Names in the computer

The public has a fear of having its names in a computer. Yet, locked up in a computer, accessible only to those with special knowledge and the right of access, names are much safer than on bits of paper. Computer encrypting techniques are now such that they are, for practical purposes, unbreakable.

Of legitimate public concern is the matching of names of cancer patients with those in other, non-medical files. The Nordic countries have data protection boards which approve such matching on a study-by-study basis. In England and Wales, the approval of the Ethics Committee of the British Medical Association is needed. Such open authorization, given after the investigator has explained his aims, is more likely to control abuse and result in better studies than surreptitious matching.

It must be stressed that absolute secrecy means in effect that the individual cancer patient is prevented from benefiting from the experience of others with cancer and from contributing to the pool of knowledge about the disease; absolute secrecy makes it easier for industrial and other risks to remain undiscovered or deliberately concealed; absolute secrecy prevents the community from assessing the value received from funds invested in screening and prevention programmes. Those who continue to oppose ethical cancer registries bear a heavy responsibility. It must be emphasized that when the results of epidemiological studies are published, names are *never* given. Nonetheless, registries must ensure that all steps are taken to ensure that the data in their care on registered persons does not reach unauthorized third parties. To this end, it is useful to have a check-list of points in the registration process which may be vulnerable and a list of measures to be observed to ensure secrecy. Legislation in Australia provides for penal sanction for deliberate breaches of confidentiality.

STAFFING AND FUNDING

A cancer registry is useful only if it is properly staffed and funded. Schraub of the Doubs Cancer Registry has observed that the funds allocated to cancer registration in this department of France in one year are about those incurred in treating three lung cancer patients—a form of malignancy with a very high mortality.

Costs per case registered vary from country to country. At one end of the scale, the SEER programme in the USA covers 10% of the US population by a system of active registration involving extracting of hospital records by trained registry staff at a cost of about US$100 per case. A considerable amount of detail is collected on each cancer patient. Averaged out, the annual cost of the entire programme is less than 45 cents per head of population. (In 1982 the costs of cancer in the USA were estimated at over US$40 billion, or US$180 per head.)

The cancer registry needs four types of staff:

(1) data collectors—those who ensure that notifications are collected and that no source is missed;

(2) bookkeepers—those who ensure that notifications are complete, that the data are consistent with any previously obtained and that duplicates are recognized, and who match registrations against death certificates and lists of cohort members;

(3) analysts—persons who prepare tables, an annual report and respond to requests for information;

(4) interpreters—those who use the registry data for epidemiological and clinical studies. Such persons are not necessarily on the staff of the registry, but may work in schools of medicine, public health or government departments. However, cancer registry staff often have a better knowledge of the validity of the registry data.

Much of the current disenchantment about cancer registration lies in failure to provide an adequate complement of analysts and interpreters so that the data collected are used. The need for an *interested* and preferably full-time director is paramount—the quality of a director is soon reflected in the work and productivity of his registry.

COMMENT

Cancer registration is but one of the activities in cancer control, but registration is likely to increase in importance as attempts to discover more about the causes of human cancer continue. Although experimental approaches are of great value for studying cancer related to exposures to single, defined chemicals, they are less useful for elucidating the complex interaction of diet and lifestyle in humans.

One might ask where the opposition to ethically conducted cancer registration comes from. The widespread concern about the privacy of the citizen, by individuals who may not have realized that absolute privacy may lead to the concealment of cancer risk, has already been mentioned. Less enlightened industrialists and governments see in the cancer registry a means of uncovering risk to their workforce that they would rather conceal. This is a singularly short-sighted policy. Risk will always emerge. The vast sums to be paid in compensation to asbestos workers in some countries arise from willful concealment. Further, without registries, it would be very difficult to show that a particular exposure is unlikely to be dangerous to health.

In order that differences in cancer patterns between ethnic or national groups become clear, such information must be recorded. Unfortunately, for political reasons, governments may not wish to recognize the presence of migrant groups within their country, and the opportunity for identifying the existence of, and the reasons for, differences in cancer risk between ethnic groups is lost. Reluctance to identify ethnic groups may prevent identification of groups in greatest need, e.g., American Indians in the USA (Young *et al.*, 1984).

The cancer registry is a valuable tool in cancer control. To be truly effective it must be possible to record the names of those with the disease and to match these, with proper safeguards, to those in other files.

REFERENCES

Armstrong, B. & Doll, R. (1975) Environmental factors and cancer incidence and mortality in different countries with special reference to dietary practices. *Int. J. Cancer, 15,* 617–631

The Cancer Registry of Norway (1980) *Survival of Cancer Patients. Cases Diagnosed in Norway 1968–1975,* Oslo

Cole, P. (1979) The evolving case-control study. *J. chronic Dis., 32,* 15–27

Commission of the European Communities (1984) *The Confidentiality of Medical Records—The Principles and Practice of Protection in a Research-dependent Environment. Medicine No. EUR9471En,* Luxembourg

Council of Europe (1981) *Convention No. 108 for the Protection of Individuals with Regard to Automatic Processing of Personal Data; Council of Europe (1981) Recommendation No. R(81)1 of the Committee of Ministers of Member States on Regulations for Automated Medical Data Banks,* Strasbourg

Day, N.E. & Boice, J.D., Jr, eds (1984) *Second Cancer in Relation to Radiation Treatment for Cervical Cancer: Results of a Cancer Registry Collaboration (IARC Scientific Publications No. 52),* Lyon, International Agency for Research on Cancer

De Vita, V.T., Simon, R.M., Hubbard, S.M., Young, R.C., Berard, C.W., Moxley, J.H., III, Frei, E., III, Carbone, P.P. & Canellos, G.P. (1980) Curability of advanced Hodgkin's disease with chemotherapy. Long-term follow-up of MOPP-treated patients at the National Cancer Institute. *Ann. intern. Med., 92,* 587–595

Doll, R. & Peto, R. (1976) Mortality in relation to smoking: 20 years' observation on male British doctors. *Br. med. J., ii,* 1525–1536

Doll, R. & Peto, R. (1981) The causes of cancer: Quantitative estimates of avoidable risks of cancer in the United States today. *J. natl Cancer Inst., 66,* 1191–1336

Doll, R., Muir, C. & Waterhouse, J. (1970) *Cancer Incidence in Five Continents. Volume II,* Berlin, Springer, pp. 38–44

Hakama, M. (1982) *Trends in the incidence of cervical cancer in the Nordic countries.* In: Magnus, K., ed., *Trends in Cancer Incidence, Causes and Practical Implications,* Washington DC, Hemisphere Publishing Corp., pp. 279–292

Heasman, M.A. & Lipworth, L. (1966) *Accuracy of Certification of Cause of Death (Studies on Medical and Population Subjects, No. 20),* London, Her Majesty's Stationery Office

Henderson, B.E., Ross, R.K. & Pike, M.C. (1983) *Exogenous hormones and the risk of cancer.* In: *Recent Advances in Cancer Control. Proceedings of the 6th Asia Pacific Cancer Conference, Sendai, Japan,* Amsterdam, Excerpta Medica, pp. 73–85

Higginson, J. & Muir, C.S. (1979) Environmental carcinogenesis: Misconceptions and limitations to cancer control. *J. natl Cancer Inst., 63,* 1291–1298

Kreyberg, L. (1952) 100 consecutive primary epithelial lung tumours. *Br. J. Cancer, 6,* 112–119

Laara, E. (1982) *Development of Cancer Morbidity in Finland up to the Year 2002. Predictions on Incidence Rates and Numbers of New Cases for Some Common Cancers Based on Analysis by Age, Period and Cohort* [in Finnish], Helsinki, Finnish Cancer Registry

Linsell, C.A. (1979) Environmental chemical carcinogens and liver cancer. *J. Toxicol. environ. Health, 5,* 183–191

Miller, A.B., ed. (1978) *Screening in Cancer (UICC Technical Report No. 40),* Geneva, International Union Against Cancer

Muir, C.S. & Démaret, E. (1982) *Legal Basis of Cancer Registration (IARC Internal Technical Report No. 82/003),* Lyon, International Agency for Research of Cancer

Muir, C.S., Choi, N.W. & Schifflers, E. (1981) *Time trends in cancer mortality in some countries. Their possible causes and significance.* In: *Medical Aspects of Mortality Statistics (Skandia International Symposia),* Stockholm, Almqvist & Wicksell, pp. 269–309

Muñoz, N., Correa, P., Cuello, C. & Duque, E. (1968) Histological types of gastric carcinoma in high- and low-risk areas. *Int. J. Cancer, 3,* 809–818

National Center for Health Statistics (1981) *The National Death Index—User's Manual (DHHS Publication No. (PHS)81-1148),* Hyattsville, MD, Office of Health Research, Statistics and Technology

Office of Population Censuses and Surveys (1981) *Report of the Advisory Committee on Cancer Registration (MB1 No. 6),* London, Her Majesty's Stationery Office

Percy, C. & Dolman, A. (1978) Comparison of the coding of death certificates related to cancer in seven countries. *Public Health Rep., 93,* 335–350

Percy, C., Stanek, E. & Gloeckler, L. (1981) Accuracy of cancer death certificates and its effect on cancer mortality statistics. *Am. J. publ. Health, 71,* 242–250

Puffer, R.R. & Wynne-Griffith, G. (1967) *Patterns of Urban Mortality (Scientific Publication No. 151),* Washington DC, Pan American Health Organization

Roginski, C. (1982) *Comparison of urban and rural incidence data.* In: Waterhouse, J., Muir, C., Shanmugaratnam, K. & Powell, J., eds, *Cancer Incidence in Five Continents Volume IV (IARC Scientific Publications No. 42),* Lyon, International Agency for Research on Cancer, pp. 636–667

Saracci, R., Simonato, L., Acheson, E.D., Andersen, A., Bertazzi, P.A., Claude, J., Charnay, N., Estève, J., Frentzel-Beyme, R.R., Gardner, M.J., Jensen, O.M., Maasing, R., Olsen, J.H., Teppo, L., Westerholm, P. & Zocchetti, C. (1984) Mortality and incidence of cancer in workers in the man made vitreous fibres producing industry: An international investigation at 13 European plants. *Br. J. ind. Med., 41,* 425–436

Teppo, L., Pukkala, E., Hakama, M., Hakulinen, T., Herva, A. & Saxén, E. (1980) Way of life and cancer incidence in Finland. *Scand. J. soc. Med.,* Suppl. 19

Wynder, E.L. & Gori, G.B. (1977) Contribution of the environment to cancer incidence: An epidemiologic exercise. *J. natl Cancer Inst., 58,* 825–832

Young, J.L., Ries, L.G. & Pollack, E.S. (1984) Cancer patient survival among ethnic groups in the United States. *J. natl Cancer Inst., 73,* 341–352

3. PLANNING AND EVALUATING PREVENTIVE MEASURES

L. TEPPO, M. HAKAMA, T. HAKULINEN, E. PUKKALA & E. SAXÉN

Finnish Cancer Registry, Liisankatu 21 B, SF-00170 Helsinki 17, Finland

This paper discusses the role of the cancer registry in planning and evaluating preventive measures, from four points of view: (1) identification of priorities, (2) search for risk factors, (3) evaluation of preventive measures and (4) health education. The paper is based primarily on experience gained at the Finnish Cancer Registry, which is population-based and nationwide and has been in operation for more than 30 years (Saxén & Teppo, 1978).

IDENTIFICATION OF PRIORITIES

Before any preventive measure is undertaken, one should know as precisely as possible the magnitude of the problem—in other words, what kind of result would be expected if successful prevention were achieved. Measures of this magnitude include incidence rates and annual numbers of cases. The latter often provide a more straightforward view of the size of the problem. Clinical findings and mortality data give some information about cancers that are common in the population, but a more precise picture can be obtained from incidence statistics produced by a population-based cancer registry. Statistics based on causes of death are never as accurate as incidence data based on original diagnoses of cancer. In addition, many cancers are cured, and patients with such cancers do not appear in mortality statistics; thus, the impact of cancer with a favourable prognosis will be underestimated in evaluations based on mortality data. Admittedly, cancers with a poor prognosis are those that should be selected as priority targets of preventive measures.

Principal cancer sites

One of the routine activities of a cancer registry is to produce statistics on the principal cancer sites. Cancers that stand high on these lists should have priority when preventive measures are considered. It is seldom useful to try to prevent rare cancers at the population level. Table 1 is a list of leading cancer sites in four different areas of the world. Only cancer registries are in a position to produce statistics of this nature.

Table 1. Leading primary sites of cancer in Finland (1971–1976), Shanghai, China (1975), Cali, Colombia (1972–1976) and Dakar, Senegal (1969–1974), by sex (Waterhouse et al., 1982)

Finland		Shanghai		Cali		Dakar	
Primary site	%	Primary site	%	Primary site	%	Primary site	%
Males							
Lung	30.6	Stomach	23.1	Skin, other	20.3	Liver	36.8
Stomach	11.8	Lung	20.1	Stomach	18.6	Skin, other	12.4
Prostate	10.7	Liver	14.6	Prostate	7.8	Lymphoma	5.6
Bladder	3.9	Oesophagus	9.7	Lung	7.6	Stomach	4.3
Pancreas	3.9	Rectum	3.5	Bladder	3.9	Soft tissues	3.8
Rectum	3.5	Bladder	2.9	Lymphoma	3.4	Bladder	3.5
Colon	3.3	Colon	2.9	Larynx	2.4	Prostate	3.5
Kidney	3.0	Nasopharynx	2.6	Myeloid leuk.	2.3	Nervous system	3.3
Nervous system	2.5	Pancreas	1.7	Nervous system	2.2	Hodgkin's dis.	2.2
Larynx	2.2	Larynx	1.7	Pancreas	2.0	Eye	2.1
All sites	100.0	All sites	100.0	All sites	100.0	All sites	100.0
Females							
Breast	21.5	Cervix uteri	14.5	Cervix uteri	21.8	Cervix uteri	21.0
Stomach	9.8	Stomach	13.2	Skin, other	17.0	Breast	14.7
Corpus uteri	6.1	Breast	12.9	Breast	12.9	Liver	13.4
Ovary	5.8	Lung	11.1	Stomach	9.5	Skin, other	10.2
Colon	5.8	Liver	5.7	Ovary	4.3	Ovary	6.0
Cervix uteri	4.5	Oesophagus	5.0	Thyroid	2.7	Lymphoma	2.7
Rectum	4.1	Thyroid	4.4	Gallbladder	2.5	Soft tissue	2.2
Pancreas	4.1	Colon	3.7	Colon	1.8	Stomach	2.2
Lung	3.4	Rectum	3.6	Nervous system	1.4	Nervous system	2.1
Nervous system	3.1	Ovary	2.7	Lymphoma	1.4	Bladder	2.0
All sites	100.0	All sites	100.0	All sites	100.0	All sites	100.0

The cancer pattern varies by age, the main types occurring in childhood being brain tumours and leukaemia, and those in early adulthood lymphomas, leukaemia, testicular tumours and thyroid cancer (females). Thereafter, cancers of the gastrointestinal tract, breast, uterus, lung and colo-rectum head the list of cancers, with cancer of the prostate being common in elderly men. This age dependence may be of importance when assessing the impact of cancer on a community.

Prevalence of cancer

Although incidence rate gives a good estimate of the relative importance of the disease in a population, prevalence (among other measures) is also relevant for planning purposes. The prevalence of cancer is represented by the number of all cancer patients in the population at risk. The number of cancer patients is usually defined as those who have been given a diagnosis of cancer and are alive at a certain point of time (Hakama et al., 1975). The prevalence thus includes, in addition to

Table 2. New cases of selected cancers diagnosed in 1977, and the number of prevalent cases at 31 December 1977, in Finland, by sex (Finnish Cancer Registry)

Males	New cases 1977	Prevalence 31 December 1977	Females	New cases 1977	Prevalence 31 December 1977
All sites	6 848	21 035	All sites	6 517	36 208
Lung	2 071	3 099	Breast	1 439	10 821
Prostate	922	2 982	Stomach	568	1 122
Stomach	688	1 387	Corpus uteri	420	4 020
Bladder	355	1 338	Ovary	415	2 228
Pancreas	265	117	Colon	413	1 505
Rectum	254	931	Rectum	282	1 282
Colon	251	887	Lung	260	406
Lymphoma	211	841	Pancreas	236	150
Kidney	196	624	Cervix uteri	233	4 121
Leukaemia	173	363	Nervous system	189	1 126
Nervous system	157	727	Leukaemia	186	342
Lip	125	1 835	Thyroid gland	160	1 178
Larynx	105	1 133	Lymphoma	158	679
Oesophagus	103	113	Gallbladder	153	111
Skin, melanoma	94	599	Kidney	152	651
Soft tissue	38	302	Skin, melanoma	135	1 024

patients and those disabled by cancer itself or by cancer treatment, healthy persons who have been cured of cancer. Prevalence figures should give some insight into the burden of any given cancer on society.

Table 2 lists the numbers of prevalent cases of cancer in Finland compared with corresponding one-year yields of incident cases of cancer. The orders in which the various cancers appear are quite different. This is due to differences in length of survival: cancers with a good prognosis occupy high positions and those with unfavourable prognoses lie far behind, even if their incidence is high. In addition, the trend in risk affects the relationship between incidence and prevalence. Calculation of prevalence figures and rates is based on data on incident cases of cancer and of follow-up data on death. Well-functioning population-based cancer registries are usually able to produce prevalence statistics.

Years lost due to cancer

The impact on society of different cancer types, measured by loss of person-years, depends on the age of the patients. For instance, on the one hand, cancer of the prostate is a disease of the elderly, so that even if a major proportion of prostatic cancers could be eliminated, the effect (measured by prolongation of life) would be small (Hakulinen & Teppo, 1976). On the other hand, if cancers of childhood or early adulthood could be eliminated, the positive effect, as measured by saved person-years (or by saved working-years), would be much higher, even if the cancer in question were not very common.

Table 3. Increases in person-years of working age (20–64 years) resulting from the hypothetical elimination of annual deaths from cancers at various sites, by the application of the theory of competing risks (averages from 1966–1970 in Finland) (Hakulinen & Teppo, 1976)

Males	No. of years	Females	No. of years
All sites	19 500	All sites	16 500
Lung	5 400	Breast	2 900
Stomach	2 300	Leukaemia	1 900
Leukaemia	2 100	Stomach	1 600
Brain	1 700	Cervix uteri	1 300
Lymphoma[a]	1 200	Ovary	1 200
Pancreas	700	Brain	1 200
Kidney	500	Lymphoma[a]	700
Colon	500	Lung	500
Skin[b]	500	Colon	500
Bone	400	Pancreas	400
Rectum	300	Kidney	400
Larynx	300	Skin[b]	300
Urinary bladder	200	Rectum	300
Oesophagus	200	Corpus uteri	300
Prostate	200	Oesophagus	200

[a] Includes Hodgkin's disease
[b] Includes melanomas; excludes basal-cell carcinomas

Table 3 shows the results of an analysis carried out at the Finnish Cancer Registry, in which the figures, based on mortality statistics for Finland, were produced by employing the theory of competing risks. Cancers detected in childhood stand relatively high in these statistics; thus, leukaemia, brain tumours and lymphomas, although not the most common cancers in Finland, would be very good targets for preventive measures, provided that a proper method were available for this purpose. Cancer of the prostate, which is second on the list of leading cancer sites in Finland (Table 2) when measured by the annual number of new cases, comes fifteenth on the list of saved working-years and no higher than seventh on the list of saved years for ages 0–89 (Hakulinen & Teppo, 1976).

Geographical analyses and cancer maps

One of the traditional functions of cancer registries is to produce cancer incidence maps (Fig. 1). The larger and more heterogeneous the area of registration (provided that there is a sizeable population living in the area), the more useful are such maps. There are two requirements for reliable calculation of regional incidence rates: the cancer registration system must cover the whole area equally and comprehensively, and population data must exist at a regional level.

The risk of incurring certain cancers may differ substantially within one country or registration area. If technical causes (such as underreporting) can be excluded and if adjustment has been carried out for obvious confounders, the differences in the

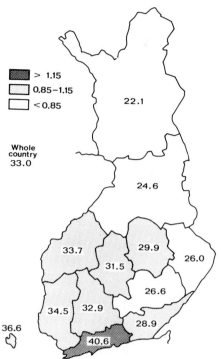

Fig. 1. Age-adjusted incidence rates of breast cancer in females in Finland in 1966–1970, by province. The stratification indicates relative rates. The north-south gradient in the rates reflects the gradients in standard of living and urbanization, which are higher in the south (Teppo et al., 1975).

incidence rates between different regions probably reflect differences in the exposure of the population to etiological factors. This information is useful when preventive measures are being planned at a regional level. Small regional differences do not, of course, warrant the direction of any action specifically at the high-risk area, but, since marked differences do exist, selective measures at a regional level can be considered.

Geographical areas, especially those based on administrative borders, are often heterogeneous. Provinces and corresponding areas usually have at least one centrally located urban district with less developed rural districts at the periphery. This means that certain differences in carcinogenic exposures that exist between, say, urban and rural districts are not reflected in the regional incidence rates. In order for such distinctions to be made, more sophisticated methodology is called for. One way to overcome this problem is to construct fictitious aggregates of small geographical units with similar background characteristics. Figure 2 gives an example of such an analysis carried out at the Finnish Cancer Registry (Teppo et al., 1980). Each of the 464 municipalities in Finland (average population, 10 000) was characterized by a number of variables related to socioeconomic status, welfare, occupations, etc. All of the municipalities were then divided into four to five classes according to the numerical

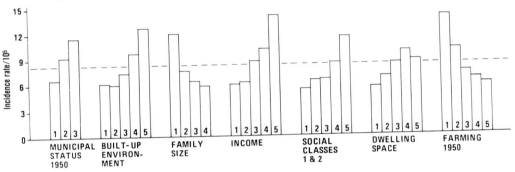

Fig. 2. Age-adjusted incidence rates of colon cancer in males in Finland in 1955–1974 in classes of municipalities defined by the value of various background variables. Horizontal dotted line: rate for the total population. Definitions of the variables and their classes: Municipal status 1950: 1 = rural, 2 = country-town, 3 = town. Built-up environment (percentage of population living in urban-like centres in 1970): 1 = lowest, 5 = highest. Family size (mean number of residents per household in 1970): 1 = lowest, 4 = highest. Income (average monthly income per inhabitant in 1968): 1 = lowest, 5 = highest. Social classes 1 & 2 (percentage of people belonging to the two highest social classes (out of four) in 1970): 1 = lowest, 5 = highest. Dwelling space (mean dwelling space per inhabitant in 1970): 1 = lowest, 5 = highest. Farming 1950 (percentage of population in farming and forestry): 1 = lowest, 5 = highest (Teppo et al., 1980)

value of each parameter. Thereafter, age-adjusted incidence rates of different cancers were calculated within each of these aggregates: the municipality and age-specific numbers of cases were added together and divided by the sum of the annual age-specific population figures for the same municipalities. It appeared that many cancers occurred at higher rates in well-developed, urbanized districts and low rates in less-developed, remote districts in which a large proportion of the population worked on the land. A similar trend was found in analyses made separately for urban and rural areas. Many other parameters can be applied and indications for the direction of preventive measures obtained. If it is not possible to do analyses like those described above, cancer registries are often able to produce simpler statistics on the incidence rates of different cancers separately for urban and rural populations (Table 4).

Cancer trends

Cross-sectional cancer statistics from one year or from a certain period of time do not reflect the entire picture of the cancer burden on society. Although the incidence rate of all cancers taken together in general does not change very much with time, many individual cancers show distinct time trends, either upward or downward (Teppo et al., 1975; Cancer Registry of Norway, 1982; Magnus, 1982). It is clear that cancers occurring with increasing incidence merit more attention than those with decreasing incidence; the decrease often continues without any active preventive measures having to be taken.

Table 4. Urban-rural ratio of age-adjusted incidence rates of selected cancers in Finland in 1975–1979 (Finnish Cancer Registry, unpublished), and in Norway in 1973–1977 (Waterhouse et al., 1982)

	Finland 1975–1979	Norway 1973–1977
Males		
Lip	0.66	0.64
Oesophagus	1.00	1.95
Stomach	0.97	1.03
Colon	1.70	1.35
Larynx	1.32	1.68
Lung	1.05	1.84
Prostate	1.20	1.16
Testis	1.29	1.12
Kidney	1.47	1.38
Skin, melanoma	0.94	1.47
Soft tissue	1.21	1.28
Multiple myeloma	1.07	0.92
All sites	1.12	1.25
Females		
Oesophagus	0.83	1.00
Stomach	1.00	1.10
Colon	1.23	1.23
Lung	1.78	1.77
Breast	1.36	1.23
Cervix uteri	1.22	1.29
Corpus uteri	1.24	0.92
Ovary	1.12	1.05
Kidney	1.32	1.14
Skin, melanoma	1.02	1.18
Thyroid	0.94	0.91
Soft tissue	1.03	0.93
Multiple myeloma	0.98	0.85
All sites	1.19	1.16

The incidence of stomach cancer among females has been decreasing in Finland over the past 30 years, whereas the incidence of cancer of the corpus uteri has recently been increasing (Fig. 3); today, these cancers occur at approximately equal incidence rates. It can be predicted that, if nothing is done, cancer of the stomach will continue to decrease in incidence and cancer of the corpus uteri will probably increase and become a real problem in years to come. Active measures aimed at eliminating corpus cancer would therefore be essential. Other cancers the incidence of which is increasing and which should therefore be considered as important targets for preventive measures in Finland are those of the colon and rectum, bladder, and melanoma in males, and those of the breast, lung, colon and rectum, melanoma and gallbladder in females (Fig. 3).

Fig. 3. Annual age-adjusted incidence rates of cancers at selected primary sites in Finland in 1953–1979, and forecasts up to the year 2000 based on a log-linear model including age, period and cohort factors, by sex (Läärä, 1982)

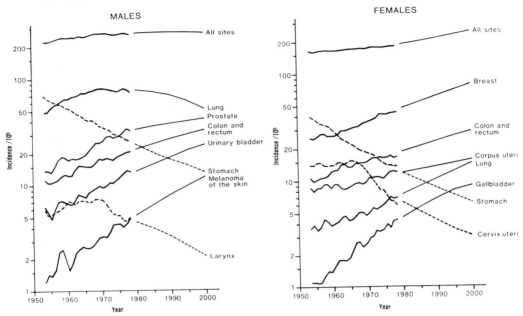

Trend analyses can be extended to geographical areas and to different age groups. In some instances, other parameters (ethnicity, religion, country of origin, etc) can also be used as a basis for determining trends. Reliable cancer incidence trends cannot be produced without a stable and well-functioning cancer registration system. However, in view of the many sources of error, such trends must always be interpreted very cautiously (Saxén, 1982).

Birth cohort analyses and predictions

Cancer incidence trends can also be used to predict future incidence trends, which —when combined with population forecasts—can be transformed to numbers of new cancer cases. Simple extrapolation is useful for short-term predictions. The forecast can be improved if three basic determinants that influence future incidence rates, i.e., age, calendar time and birth cohort, are taken into consideration (Läärä, 1982).

Figure 4 shows a birth-cohort analysis of cancer of the stomach in males and melanoma of the skin in females in Finland. The age-specific incidence of gastric cancer of each successive birth cohort is lower than that of the previous one. Thus, when younger cohorts replace the present old ones, the incidence of cancer of the stomach will certainly be much lower than it is today. For cutaneous melanoma the situation is just the opposite: each cohort has a higher rate than the previous one, and a distinct increase in the total rate is to be expected. Similar conclusions can be

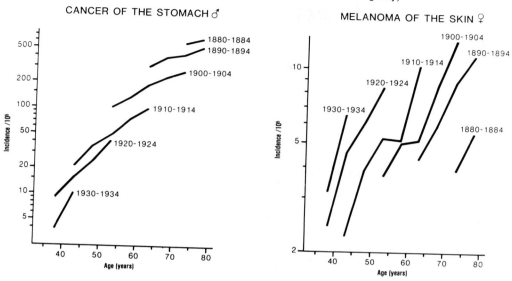

Fig. 4. Age-specific incidence rates of cancer of the stomach in males and melanoma of the skin in females in Finland, by five-year birth cohort (Finnish Cancer Registry)

obtained simply by examining the trends in the age-adjusted incidence rate (shown in Fig. 3), but birth cohort analysis improves the reliability of the forecast. In some instances, a simple trend in the total incidence shows no change, while cohort-specific trends may show that future changes are to be expected. Figure 3 shows the trend forecast for selected cancer types in Finland up to the year 2000 (Läärä, 1982).

The forecast can sometimes be improved if there are strong risk factors with high prevalence and if the trends in these risk factors are known, since there is a latent period extending to decades between the start of the exposure and the diagnosis of cancer (Hakama, 1980). Therefore, whether risk factors were taken into consideration or not, forecasts extending up to some 20 years into the future are still realistic. However, the successful removal of carcinogens with a promoting effect is likely to change the trend.

SEARCH FOR RISK FACTORS IN CANCER

It is not enough to know the magnitude of the cancer problem and to assess the priorities for action: methods should also be available for the preventive measures to be undertaken. Since this calls for knowledge of the risk factors involved, the search for risk factors is one of the main objects of cancer epidemiology, and cancer registries play an important role in this. Several approaches can be applied.

Clusters and multiple cancers

Cancer registries are often in a position to discover clusters of cancer cases. These are usually geographical, i.e., cases of cancer, preferably of one and the same type, appear within a limited geographical area more often than would be expected. Such clusters may give clues to the etiology of the cancer in question. The search for exposure clusters may also be useful; e.g., an increased risk of leukaemia was seen in children who were born during the first five months after a four-month epidemic of 'Asian' influenza in 1956 in Finland (Hakulinen *et al.*, 1973). No such association was detected between other influenza epidemics and childhood leukaemia.

Studies on multiple cancers are one way of identifying risk factors in cancer (Schoenberg, 1977). Two or more cancers may have a similar etiology, and patients with one are at an increased risk of contracting another. Cancers of the lip, larynx and lung provide such an example (Lindqvist *et al.*, 1979). In these instances, an excess risk should be observable in both directions. Sometimes different carcinogenic exposures tend to be prevalent in a certain section of the population and to lead to pairs of mutual excess risks. Both cancer of the breast and cancer of the colon are diseases of higher socioeconomic classes, although their risk factors are to a great extent different. Finally, the carcinogenic effects of cancer treatment can be analysed using cancer registry material, provided that accurate enough data on the treatment are available.

Ecological studies

As mentioned above, cancer registries are in a position to produce incidence rates by geographical area. If the same area can be characterized by a numerical value for the occurrence of a suspected risk factor, the association between this variable and the incidence rate can be analysed. A number of correlation studies have been done using countries as statistical units. Many have addressed dietary factors (Armstrong & Doll, 1975).

There are some difficulties in interpreting the results of ecological correlation studies. The populations are heterogeneous within the units, and figures that are expected to describe the exposure of the population do not always do so adequately. For instance, the prevalence of smoking is not the only parameter necessary to describe the smoking habits of a population. A positive association between a suspected risk factor and cancer incidence may also be noncausal and due to other factors associated with the risk factor under study. However, correlation studies are often useful starting points for more detailed investigations at the level of individuals.

Follow-up studies

In many follow-up studies, cancer deaths are used as the end point; however, the use of incident cases of cancer has several advantages: higher numbers of observed cases, a shorter latent period from the onset of the exposure, and reliable data on the primary site of cancer. Hence, the data base of cancer registries may serve as a useful source of information for follow-up (cohort) studies. In the Nordic countries and some other areas, linkage between a list of exposed individuals and cancer registry files can be established by computer, using the unique personal identification numbers of the individuals concerned. Since the expected numbers of cases are calculated from the same data base as the observed numbers of cases, the comparability of the two figures is greatly improved (possible underreporting reduces both the observed and expected figures in a similar way).

It is important that the data on a basic study cohort be accurate, i.e., that an individual within the cohort can be reliably identified in the cancer registry files. In order to calculate person-years at risk, it is also necessary to follow up the cohort for deaths and to identify losses from follow-up. Several types of follow-up study, involving occupational exposures (Jensen, this volume), cancer treatment (e.g., radiation: The International Radiation Study Group, 1983), other diseases (Kvåle et al., 1979), vaccination (Snider et al., 1978) and personal habits (Hakulinen et al., 1974), have been carried out successfully using cancer registry data.

Case-control studies

Case-control studies can also be done on the basis of cancer registry data (Aromaa et al., 1976). Especially in the case of rare tumours, it is seldom possible to obtain a sizeable enough study group without a cancer registry. Possible underreporting does not seriously affect the reliability of the results obtained. Sometimes, the controls can also be drawn from cancer registry files (Lindqvist, 1979).

EVALUATION OF PREVENTIVE MEASURES

Simulation models

If the prevalence of an exposure and the quantitative relationships between the carcinogenic exposure and the effect (i.e., cancer) are known, it is possible to assess —quantitatively as well as qualitatively—what would happen if a reduction of a certain magnitude were achieved in the exposure. Unfortunately, very few such relationships are known well enough to be useful in such calculations. Smoking and lung cancer provide a good, if not the only, example. The prevalence of smoking in different population groups and the risks associated with different smoking categories defined by the amount of tobacco smoked are known, as are also, to some extent, the latent periods and the effect of the age at which smoking started.

A simulation study was carried out by the Finnish Cancer Registry (Hakulinen & Pukkala, 1981). A fictitious population, similar in age structure and smoking habits to those prevalent in Finland, was constructed in a computer. The percentage of those who started or stopped smoking was permitted to vary. As a result, a number of curves were produced describing future trends in the age-adjusted incidence rate of lung cancer in males. Figure 5 shows the trend up to the year 2000, assuming that the proportion of those who start smoking remains constant in each age group and that the proportion of those who stop is 0%, 10% or 20% in each smoking category in each coming five-year calendar period. If nobody were to stop smoking in the future, the curve would rise markedly, and an increase would soon begin. If the proportion of those stopping were as high as 20%, a substantial decrease in the

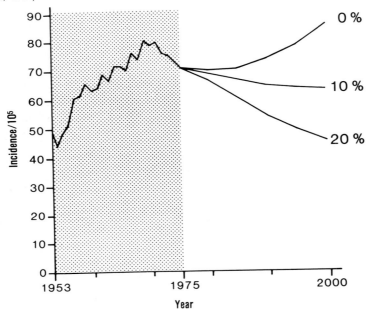

Fig. 5. Age-adjusted incidence rates of lung cancer in males in Finland in 1953–1975, and three forecasts for the rates in 1980–2000, derived from a simulation model based on the following assumptions: in each consecutive five-year period in 1976–2000, 30% of non-smokers aged 10–14 years, 15% of those aged 15–19 and 5% of those aged 20–24 will start smoking, and 0%, 10% and 20%, respectively, of smokers in each category will stop smoking (Hakulinen & Pukkala, 1981).

incidence rate would be achieved. The curves for different stopping frequencies between 0% and 20% can be interpolated easily.

One can also keep the frequency of stopping in each smoking category constant and let the frequency of starting smoking vary. Figure 6 gives three alternatives, each composed of different combinations of starting frequencies. Curve 2 shows what would happen if the current frequencies were to remain unchanged. Because of the latent period, the effect of starting smoking does not appear for 25 years, before the year 2000; however, a wide variation in the effect is achieved. It is clear that the parameters included in the model act simultaneously and that there are other parameters, which have not been or cannot be taken into account. However, this kind of simulation study, based on cancer incidence rates in a population, reveals estimates of the magnitude of the effect that can be reached.

Evaluation of primary prevention (monitoring)

If specific preventive measures to reduce the incidence of a certain type of cancer have been undertaken (which means that the risk factors are known and that there are methods to eliminate them or to reduce the exposure to them), it will be necessary sooner or later to establish whether the effort has had any positive effect. One way

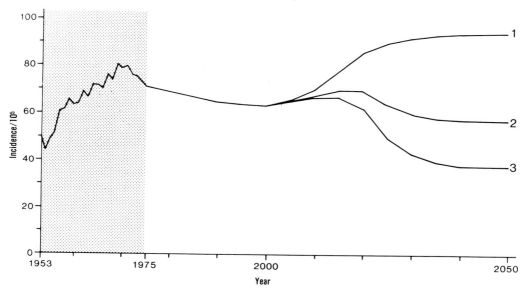

Fig. 6. Age-adjusted incidence rates of lung cancer in males in Finland in 1953–1975, and forecasts for the rates in 1980–2050 derived from a simulation model based on the following assumptions: in each consecutive five-year period in 1976–2050, 10% of smokers in each smoking category will stop, and one of the following three alternatives (three curves) holds true: (1) 60% of non-smokers aged 10–14 years, 30% of those aged 15–19 and 10% of those aged 20–24 will start smoking; (2) the percentages are 30, 15 and 5, respectively; and (3) the percentages are 15, 7.5 and 2.5, respectively (Hakulinen & Pukkala, 1981).

is to monitor the exposure, e.g., the prevalence of smoking. A positive effect indicates that, technically at least, the action taken is working. However, this is not a sufficient or adequate measure of the success of a system: in addition to information on the exposure, information on the disease itself should also be obtained. In cancer, incidence figures produced by cancer registries are the parameters usually monitored (Muir et al., 1976). Incidence trends may show that the risk has started to decrease, which suggests that the action has been successful, or the trend may remain unchanged, which is likely to indicate failure.

At least three points should be kept in mind: (1) It is important also to have data on incidence trends before the preventive measures were instituted. If a curve has shown a downward trend for a long period of time, there is no justification for attributing a recent decrease to the specific preventive measure in question. (2) It is usually many years or even decades before any effect becomes apparent; hence, final conclusions should not be drawn too soon. (3) If the measures undertaken are directed at only a fraction of the population defined by, say, age, place of residence or occupation, the evaluation should be limited to the same subpopulation; country-wide curves do not reflect the true situation if only a limited area was subjected to the action. This does not mean that preventive activities should necessarily cover the whole population; on the contrary, it is often useful to start with specific measures within a section of the population and consider the rest as a control. Possible

Fig. 7. Trends in the proportion of current smokers among the adult population of Finland, by sex (Teppo, 1984). The arrow indicates passage of the anti-smoking law.

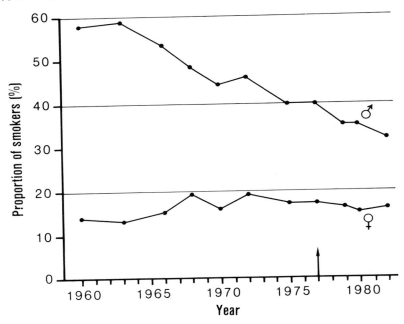

differences in incidence curves for the study population and for the control population may provide stronger evidence of a positive effect of intervention than a single curve, which is always influenced simultaneously by many other factors than the preventive measure in question.

In Finland, the incidence of lung cancer in males increased steadily from the early 1950s. In the early 1970s, the curve levelled off and has not since reached the limit of 80 per 100 000 population (Teppo, 1984); on the contrary, a slight decrease has taken place. In females, the trend has been towards a steady increase (Fig. 3).

The trend in the proportion of current smokers among adults in Finland is shown in Figure 7. The proportion of male smokers decreased consistently between the early 1960s and the late 1970s. Measured in this way, the effect among males of activities aimed at reducing smoking must be considered to have been effective. The decreasing trend in the proportion of smokers agrees well with the incidence curve for male lung cancer, although it took some 10 to 20 years before the cancer registry could confirm the favourable effect of anti-smoking activities. The frequency of smoking in females has remained relatively stable over the last two decades; previously, the percentage was lower. The incidence of lung cancer among females is still increasing, and no dramatic change in the trend of the incidence curve is to be expected in the near future.

A law was passed in Finland in 1977 with the aim of reducing smoking and, subsequently, the impact of smoking-related diseases on the population. The main provisions of this law are:

(1) Advertising and sales promotion of tobacco are prohibited.
(2) Smoking is prohibited in all public places and public transport facilities except in separated areas.
(3) The sale of tobacco products to people under 16 years of age is prohibited.
(4) The Government is responsible for fixing the upper limits of harmful components allowed in tobacco products.

The law has had no noticeable influence on the frequency of smoking among adults (Fig. 7). The effects of a reduction in the tar content of cigarettes and the effect of the law on the smoking habits of young people, and other results, either do not show up in the curve of the prevalence of smoking or will appear later. No effect can yet be expected in the trend of the long-term adverse effects of smoking, such as the incidence of lung cancer.

It is not easy to construct a useful monitoring system within a cancer registry. A simple trend in an age-adjusted incidence rate may give some indication of the success or failure of the measures undertaken. However, one should probably monitor specific age groups, specific histological types, and possibly even certain geographical areas. Thus, the number of curves to be followed up increases substantially, the statistical power is reduced, and significant changes will appear by chance.

Evaluation of screening activities

If preclinical stages of cancer are subject to mass screening, the programme can be evaluated by monitoring the incidence of the (clinical) cancer itself. A cancer registry can provide data on cases of cancer diagnosed after a single trial screening or between screenings, for the total population and sometimes separately for those attending and not attending a mass screening, which improves the power of monitoring. A good example is provided by screening of cervical cancer and its preclinical stages. This subject is covered more thoroughly in the paper by Parkin and Day in this volume. The effect of screening symptomless cancers (e.g., breast cancer by mammography) cannot be evaluated by monitoring the incidence of the cancer; what should be monitored is mortality from the cancer in question among the population subjected to screening.

A cancer registry can also establish high-risk groups characterized e.g., by attendance status, and suggest special measures, in addition to normal screening practice, for these groups (Hakama & Pukkala, 1977).

Information on the risk of cancer in a screened population and in a non-screened population and follow-up of people who have attended a mass screening may be used to draw conclusions about the natural history of different precancerous lesions and cancer (Hakama & Räsänen-Virtanen, 1976). It is possible to study how often the precancerous lesions progress into clinical cancer and how often the development of the tumour is slow enough to permit it to be detected at a preclinical stage.

It must be stressed that a cancer registry can make a reliable evaluation only if data can be linked at the level of the individual. A cancer registry should be able to identify those who attended a screening and those who died of cancer.

At present, the only screening programmes conducted routinely in the developed world are those for cervical cancer. Most vary considerably in their content and must be evaluated individually. Mass screening for the great majority of cancers is still at the experimental or planning stage. If the task of evaluating mass screening programmes is given to a cancer registry, it will not be necessary to build up a new organization for this purpose each time.

CANCER REGISTRIES AND HEALTH EDUCATION

Health education is an important tool in activities directed at the prevention of cancer. Several types of data that are produced by cancer registries can be used in health education, including information on the magnitude of the cancer problem, past and future trends in incidence and in the numbers of cases of different cancers, survival figures based on unselected patient material, geographical distribution of cancer within the registration area, etc. Moreover, depending on the local conditions, cancer registry staff may themselves be in a position to act as experts in cancer epidemiology and cancer prevention by participating in working groups and on expert committees, submitting expert opinions to administrators, etc. The participation of a cancer registry in health education depends on the interest and ability of its staff. A competent staff with epidemiological, statistical, data processing and medical expertise is a prerequisite for a well-functioning cancer registry.

If a cancer registry is reliable and its activities well known, it will continuously receive various kinds of inquiries. They may be for simple patient statistics, usually for local use (hospital district, town, municipality), for estimates of patient survival, for data on the occurrence of cancer worldwide, and other uses. Such inquiries are made by scientists, journalists, administrators, factory doctors, labour union representatives and many others. If the registry can meet the needs of society by functioning as an active research institute, it will certainly promote the prevention of cancer in the long run.

REFERENCES

Armstrong, B. & Doll, R. (1975) Environmental factors and cancer incidence and mortality in different countries, with special reference to dietary practices. *Int. J. Cancer,* **15,** 617–631

Aromaa, A., Hakama, M., Hakulinen, T., Saxén, E., Teppo, L. & Idänpään-Heikkilä, J. (1976) Breast cancer and use of rauwolfia and other antihypertensive agents in hypertensive patients: A nationwide case-control study in Finland. *Int. J. Cancer,* **18,** 727–738

The Cancer Registry of Norway (1982) *Trends in Cancer Incidence in Norway 1955–1978,* Oslo

Hakama, M. (1980) Projection of cancer incidence: Experiences and some results in Finland. *World Health Stat. Q.,* **33,** 228–240

Hakama, M. & Pukkala, E. (1977) Selective screening for cervical cancer. Experience of Finnish mass screening system. *Br. J. prev. soc. Med.,* **31,** 238–244

Hakama, M. & Räsänen-Virtanen, U. (1976) Effect of a mass screening program on the risk of cervical cancer. *Am. J. Epidemiol.,* **103,** 512–517

Hakama, M., Hakulinen, T., Teppo, L. & Saxén, E. (1975) Incidence, mortality or prevalence as indicators of the cancer problem. *Cancer, 36,* 2227–2231

Hakulinen, T. & Pukkala, E. (1981) Future incidence of lung cancer: Forecasts based on hypothetical changes in the smoking habits of males. *Int. J. Epidemiol., 10,* 233–240

Hakulinen, T. & Teppo, L. (1976) The increase in working years due to elimination of cancer as a cause of death. *Int. J. Cancer, 17,* 429–435

Hakulinen, T., Hovi, L., Karkinen-Jääskeläinen, M., Penttinen, K. & Saxén, L. (1973) Association between influenza during pregnancy and childhood leukaemia. *Br. med. J., iv,* 265–267

Hakulinen, T., Lehtimäki, L., Lehtonen, M. & Teppo, L. (1974) Cancer morbidity among two male cohorts with increased alcohol consumption in Finland. *J. natl Cancer Inst., 52,* 1711–1714

The International Radiation Study Group on Cervical Cancer (1983) *Summary chapter*. In: Day, N.E. & Boice, J.D., Jr, eds, *Second Cancer in Relation to Radiation Treatment for Cervical Cancer (IARC Scientific Publications No. 52),* Lyon, International Agency for Research on Cancer, pp. 137–181

Kvåle, G., Høiby, E.A. & Pedersen, E. (1979) Hodgkin's disease in patients with previous infectious mononucleosis. *Int. J. Cancer, 23,* 593–597

Läärä, E. (1982) *Development of Cancer Morbidity in Finland up to the Year 2002* [in Finnish] *(Health Education Series, Original Reports 3/1982),* Helsinki, National Board of Health

Lindqvist, C. (1979) Risk factors in lip cancer: A questionnaire study. *Am. J. Epidemiol., 109,* 521–530

Lindqvist, C., Pukkala, E. & Teppo, L. (1979) Second primary cancers in patients with carcinoma of the lip. *Commun. Dent. oral Epidemiol., 7,* 233–238

Magnus, K., ed. (1982) *Trends in Cancer Incidence. Causes and Practical Implications,* New York, Hemisphere Publishing Corp.

Muir, C.S., MacLennan, R., Waterhouse, J.A.H. & Magnus, K. (1976) *Feasibility of monitoring populations to detect environmental carcinogens*. In: Rosenfeld, C. & Davis, W., eds, *Environmental Pollution and Carcinogenic Risks (IARC Scientific Publications No. 13/INSERM Symposium Series Vol. 52),* Lyon, International Agency for Research on Cancer, pp. 279–293

Saxén, E. (1982) *Trends: Facts or fallacy*. In: Magnus, K., ed., *Trends in Cancer Incidence. Causes and Practical Implications,* New York, Hemisphere Publishing Corp., pp. 5–16

Saxén, E. & Teppo, L. (1978) *Finnish Cancer Registry 1952–1977. Twenty-five Years of a Nationwide Cancer Registry,* Helsinki, Finnish Cancer Registry

Schoenberg, B.S. (1977) Multiple primary malignant neoplasms. The Connecticut experience, 1935–1964. *Recent Res. Cancer Res., 58*

Snider, D.E., Comstock, G.W., Isidro, M. & Caras, G.J. (1978) Efficacy of BCG vaccination in prevention of cancer: An update: Brief communication. *J. natl Cancer Inst., 60,* 785–788

Teppo, L. (1984) *Lung cancer in Scandinavia: Time trends and smoking habits*. In: Mizell, M. & Correa, P., eds, *Lung Cancer: Causes and Prevention,* Deerfield Beach, Chemie International, pp. 21–31

Teppo, L., Hakama, M., Hakulinen, T., Lehtonen, M. & Saxén, E. (1975) Cancer in Finland 1953–1970: Incidence, mortality, prevalence. *Acta pathol. microbiol. scand. Sect. A,* Suppl. 252

Teppo, L., Pukkala, E., Hakama, M., Hakulinen, T., Herva, A. & Saxén, E. (1980) Way of life and cancer incidence in Finland. A municipality-based ecological analysis. *Scand. J. soc. Med.,* Suppl. 19

Waterhouse, J., Muir, C., Shanmugaratnam, K. & Powell, J., eds (1982) *Cancer Incidence in Five Continents Vol. IV (IARC Scientific Publications No. 42),* Lyon, International Agency for Research on Cancer

4. EVALUATING AND PLANNING SCREENING PROGRAMMES

D.M. PARKIN & N.E. DAY

International Agency for Research on Cancer, Lyon, France

INTRODUCTION

'Secondary prevention' is the term frequently used for procedures that are designed to identify cancers at an early stage, when the results of treatment may be more successful in preventing manifestations of late disease, or death. One method used to achieve such early recognition is screening—the application of simple tests or procedures that classify the population examined into individuals who probably do, and those who probably do not, have the disease in question. Screening tests are not intended to be diagnostic; positive findings must be followed up with appropriate diagnostic procedures (Wilson & Junger, 1968). The usual objective of cancer screening is to identify tumours at an early, asymptomatic stage; since treatment in the early stages of disease is likely to be more successful in reducing complications or preventing death, the rationale for screening appears to be quite clear. However, there are problems in evaluating how effective such screening programmes actually are, as will be described. For some cancers (e.g., of the oral cavity and uterine cervix), the screening procedures aim to identify stages in the natural history of the disease *prior* to the onset of invasion. In such cases, the objective is to prevent the occurrence of invasive disease, as well as its complications (including death).

The decision to implement a screening programme must be based upon good evidence that application of the procedure to a population in a particular way can reduce cancer mortality (and perhaps morbidity). The choice of programme depends on a complex series of considerations, including the nature of the population and the natural history of disease in it, and the health care services that are available. Successful planning of health care programmes is a cyclical process requiring continuous evaluation of the results achieved, so that the effectiveness of the system can be judged, and changes introduced (Knox, 1979).

THE EFFECTIVENESS OF SCREENING PROCEDURES

Most cancer screening aims to detect early preclinical disease, and the fact that early cancers have a better prognosis than advanced disease has an intuitive appeal as evidence of effectiveness. Thus, it has frequently been demonstrated that cases of lung cancer detected by chest X-ray screening are likely to be in a less advanced stage and have longer duration of survival than cases diagnosed conventionally as a result of symptoms. Since, however, randomized trials of the effectiveness of screening for lung cancer have demonstrated no reduction in mortality from the disease, it is clear that there are problems in interpreting this type of result (Taylor et al., 1981). Firstly, if the people undergoing screening are not a randomly selected group, some selection bias is likely, in that people who are screened are likely to be health-conscious individuals who, even in the absence of screening, may have had a better prognosis than a group that did not accept such tests. Even with random allocation, however, allowance has to be made for the phenomena of lead time and length bias. 'Lead time' is the interval between the time of detection by screening and time at which the disease would otherwise have been diagnosed; calculations of gains in survival must allow for this period by which diagnosis has been advanced (Hutchinson & Shapiro, 1968). 'Length bias' arises because screening tests that are applied to a population at intervals are more likely to detect cancer cases with a long preclinical phase than faster-growing tumours. Cases detected by screening are thus likely to be a biased sample of all cases, with a relatively favourable prognosis (Feinleib & Zelen, 1969). When the condition sought by screening is a precursor of invasive cancer (as in cervical cancer screening), interpretation of results of screening programmes is additionally complicated by the fact that many of the lesions discovered would never have led to cancer, so that their discovery and treatment is of much less benefit to the population than might be supposed.

For these reasons, an appraisal of the effectiveness of screening can be made only by measuring to what degree the objectives of screening are achieved. When screening for early cancer, these objectives are reductions in mortality (and, perhaps, late complications of disease); when screening for precancer, the objective is a reduction in incidence of invasive disease. The most satisfactory way of performing such an evaluation is by means of a randomized controlled trial of screening procedures. However, such trials of complex interventions, involving repeated examinations and follow-up of large numbers of individuals, are difficult to organize, expensive and time consuming. It is generally possible to investigate only a limited number of screening schedules, and the methodology used may be deemed obsolete or inappropriate by the time results become available. Finally, the apparently obvious benefit to be gained from early diagnosis means that it is difficult to convince physicians and their patients of the need for a controlled trial, in which a proportion of the population is not offered the screening programme; the result is that such trials may be judged unethical, or the control group may receive screening outside the trial.

For these reasons, evaluation of the effectiveness of screening for most sites of cancer has been based on the data from non-randomized trials of screening, and from the results of screening programmes introduced directly as a service.

Evaluation of screening for cervical cancer

The observation that cytological examination of material from the cervix uteri is a simple method for identifying potentially precancerous lesions led to the widespread introduction of the 'Pap test' as a screening procedure for carcinoma of the cervix (World Health Organization, 1969). No randomized trial of the effectiveness of cytological screening in preventing clinical invasive cancer has been performed, and there has been controversy over its true value (Foltz & Kelsey, 1980). Nevertheless, an enormous amount of data results from screening programmes which provide indirect evidence of their effectiveness; and the role of cancer registration in documenting incidence rates of disease has been of major importance.

Descriptive studies: Numerous studies have been carried out of registration rates (incidence) of invasive cervical cancer in areas in which screening has been introduced. A certain amount of care has to be exercised in interpreting such data. Firstly, it is possible that cases of carcinoma *in situ* may be erroneously registered as invasive cancer. In areas with active screening programmes, many in-situ cases are identified; and since current information on the natural history of such lesions (Boyes *et al.,* 1982) suggests that perhaps only half would have become invasive, the incidence rate derived from cancer registration can be considerably inflated. In addition, however, cytological screening brings to light early invasive cancers (micro-invasive, occult invasive). Although most such cases would have been clinically diagnosed at a later date, so that the cumulative incidence of registered cases remains unchanged, the introduction of screening may be followed by an apparent *increase* in incidence as prevalent sub-clinical cases are detected, before a fall is observed.

A more general criticism of time-series data is, however, that in the absence of any control group, it is impossible to know what would have happened in the absence of screening. Thus, a decline in incidence in a population undergoing screening may represent a general fall in risk of disease, rather than a beneficial effect of the screening.

Data from the Connecticut Tumor Registry were used to study changes in the incidence of cervical cancer in relation to cytological screening in the State (Laskey *et al.,* 1976). The age-adjusted incidence of invasive cancer fell 40% in the twenty-year period 1950/54 to 1970/73, while the number of registrations of in-situ cancer rose by almost nine fold. In two other studies—in Olmsted County, Minnesota (Dickinson *et al.,* 1972) and in Louisville, Kentucky (Christopherson *et al.,* 1976)—special registers were created to record all newly diagnosed in-situ and invasive cancers; in both areas, a decline in incidence of invasive cancer accompanied an active screening programme. However, interpretation of trends in the USA is obscured by the fact that, in general, rates were declining before any screening had been introduced (although this does not appear to have been the case in the Olmsted County study).

The Walton Report (1976) presented data from several Canadian provinces. In the three prairie provinces, Alberta, Saskatchewan and Manitoba, a general decline in incidence in the middle age range had accompanied the introduction of screening, although only in the former two provinces were data available for a period before screening was introduced, and in Alberta incidence appeared to be falling even then. Subsequent analyses suggest that the fall in incidence prior to 1972 has slowed down, especially in the younger age groups where there has been a corresponding increase in the number of registrations of carcinoma *in situ* (Canadian Task Force on Cervical Cancer Screening Programs, 1982). It was considered possible that, were it not for the screening programmes, the registered incidence of invasive cancer might have risen.

In British Columbia the provincial laboratory maintains a special register of new cases of carcinoma of the cervix (Fidler et al., 1968); incidence rates of clinical invasive disease are recorded separately from in-situ and preclinical invasive cases. Although pre-screening rates are not available, the incidence of clinical disease shows a marked decline, although the rate of fall has slowed since about 1972 (Boyes et al., 1981).

Several studies have related the intensity of screening in different areas with corresponding changes in incidence or mortality rates from cervix cancer. Cramer (1974) found a positive correlation between the decline in mortality rates and the level of screening in the USA. Miller et al. (1976) related an index of screening intensity (smears/1000 women per year) with changes in mortality from uterine cancer for Canadian provinces and for counties or census subdivisions. The results suggest that greater levels of screening are associated with larger declines in mortality, even when controlling for various socio-demographic variables of the geographic units. Lynge (1983) analysed data on cumulative incidence in women from 30–59 years of age and intensity of screening for counties of Denmark. A considerable decline in incidence was seen in counties in which organized screening programmes had been introduced in the 1960s, smaller declines being seen in areas without organized programmes but with higher or equivalent levels of screening activity.

The most impressive use of data on incidence in relation to screening is that from the Nordic countries (Hakama, 1982), where population-based registries have been in operation for serveral decades. The screening experience of these countries has been rather different; Figure 1 shows time trends in incidence. The fall in incidence is closely related to the coverage offered by the organized mass screening programmes. In both Finland and Iceland, nationwide programmes were introduced in the early 1960s, but the Icelandic scheme covers a wider age-group (25–70 years compared to 30–59) and personal invitations are sent to participants at two- to three-year intervals (five-yearly in Finland). In Sweden, the programme was introduced rather more gradually, county by county, while in Denmark only about 40% of the population live in counties offering an organized mass screening service. In Norway, the programme introduced in Ostfold county as an experimental project (Pedersen et al., 1971) has been the only mass programme available.

A study of time trends in age-specific rates in these countries shows that the falls in incidence are confined almost entirely to the age groups that have been subjected to screening, and there has been little change or even an increase in incidence in women aged 60 or more in all of the countries.

Trends in incidence rates by stage of disease have been published for Iceland (Johannesson et al., 1982), and are shown in Figure 2 along with mortality rates. Incidence of early disease (both stage IA + IB) rose in the first years of screening, but has since fallen. This illustrates the effects of screening in detecting early invasive lesions and in advancing the time and age of diagnosis over those that would have been observed without screening. The mortality rate and incidence of advanced disease have fallen in parallel, with a 60% reduction over a 10–15-year period. Given the trends before screening was introduced and the rapidity with which these falls occurred, it seems reasonable to attribute most of the reduction in risk to the effect of screening.

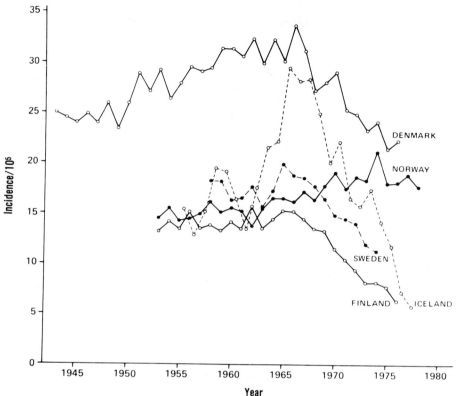

Fig. 1. Annual incidence of cervical cancer in the Nordic countries between 1943 and 1978 (from Hakama, 1982)

Recent data from the British Columbia provincial cancer registry (Miller, 1982) suggest that since about 1971 there has been an increase in the incidence of invasive cancer in women under the age of 35. This has been accompanied by an impressive rise in registrations of carcinoma *in situ,* despite the fact that the population has been well screened for many years. A similar phenomenon has been observed in recent years in England and Wales (Draper & Cook, 1983): incidence rates of invasive cancer for age groups under 35 have doubled at the same time as there have been large increases in the numbers of registrations of carcinoma *in situ* in young women. In the age group 25–34, the annual number of smears taken annually in England and Wales has increased only slightly, the increased incidence of carcinoma *in situ* being due to sharp increases in the rates of disease per 1000 smears in young women (Roberts, 1982). It seems probable that there has been a considerable increase in risk of cervical cancer in young women, and that screening is holding in check a much larger potential increase in invasive disease than has actually been observed. Parkin *et al.* (1985) have used registration data on in-situ and invasive cancers to estimate what the incidence

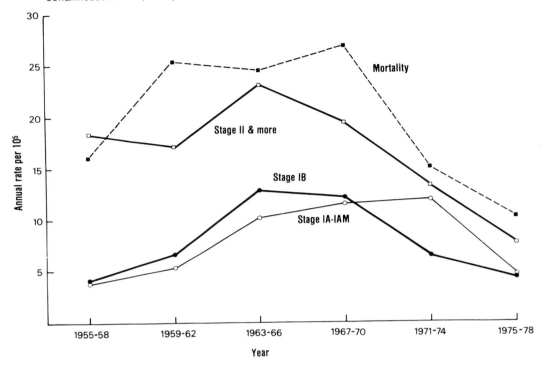

Fig. 2. Changes over time in mortality and incidence of cervical cancer. The incidence is given by stage of disease. The rates are average annual age-specific rates, in the age range 20–75 (from Johannesson et al., 1982)

rates *might* have been in the absence of screening, using a simple model which involves assumptions about the lead times of registered in-situ cases. The use of such a model is, in effect, an attempt to circumvent the defect of time-series studies by creating a theoretical unscreened control group. When incidence rates for England and Wales are plotted for different birth cohorts, the observed rates are very similar to those of Denmark and Sweden (Hakama, 1982); in all three countries there has been moderate screening activity since the mid-1960s. However, the curves 'expected' in the absence of screening closely resemble those for the data from Norway, where, as described, there has been little organized screening (Fig. 3).

Analytical studies: The data in the previous section refer to population measures of screening activity and cancer risk. More direct evidence of an effect of screening is obtained by relating cancer risk to screening activity at the individual level—that is, comparing cancer mortality and incidence rates among groups of women with different screening histories. Cohort studies have been reported from Iceland and British Columbia, areas where centralized records make this approach possible. In Iceland (Johannesson *et al.*, 1982), it was found that the mortality rate and incidence of advanced tumours were low among women who had had at least one negative

Fig. 3. Comparison of observed and theoretical incidence rates for birth cohorts in England and Wales with incidence in Norway. OBS, Rate of invasive cervix cancer observed in England and Wales; ADJ, theoretical rate obtained by the addition of half of the in-situ cancers detected by screening (from Parkin et al., 1985)

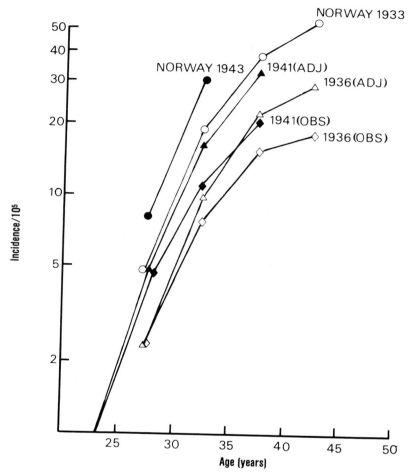

smear and were almost zero after two negative tests; rates in the unscreened group were slightly higher than before the introduction of screening, so that the relative risk compared to the screened group was around 10. Fidler et al. (1968) estimated the 'ever-screened' and 'never-screened' populations in British Columbia and calculated an age-adjusted relative risk for clinical carcinoma of 6.8 in the unscreened compared to the screened group. The screened group excluded women who developed cancer within 10 months of entering the programme (in order to exclude those cases in which the smear was taken as part of an investigation of a symptomatic tumour). Some differences may derive from the lack of comparability of the two groups. It is known, for instance, that certain risk factors for cervical cancer are associated with prior attendance at screening programmes, e.g., social class, marital status (Parkin *et al.*, 1981). A ten-fold difference in risk, however, cannot plausibly be ascribed to confounding (Bross, 1967; Breslow & Day, 1980). Furthermore, in Iceland, the unscreened group constituted about 10% of the adult female population and had a mortality rate only marginally higher than that of the general population before screening started. Self-selection for screening would have had at most a minor effect. In British Columbia, Fraser and Boyes (1976) suggested that the *screened* group contains an excess of women of lower social class and a deficit of the sexually inactive. Furthermore, since the proportion of the population in the screened group increased from 25% to 85% between 1961 and 1973, and assuming that women at higher risk of cancer entered later, the constant rate of invasive cancer in the screened population would not have been observed in the absence of a screening effect (Guzick, 1978).

A case-control approach offers an alternative method of obtaining essentially equivalent results. The rationale behind this type of study has received some attention recently (Morrison, 1982; Weiss, 1983). Essentially, it is necessary to compare the screening history of patients with clinically diagnosed invasive cancer with that in a control group without the disease. There are some methodological problems peculiar to case-control studies of the screening process. Principal among these is the need to record screening examinations prior to disease diagnosis, and not to include tests done for diagnostic purposes (which would bias the results, since it is the value of smears as screening tests, which by definition seek asymptomatic disease, and not as a diagnostic investigation which is being evaluated). For controls, the screening history will comprise all tests performed up to the time of diagnosis of the case.

This type of study has several advantages. First, screening histories are not required for an entire population of women, but only for the cancer cases (or deaths) and a corresponding control group. By limiting the study to a relatively small number of individuals, collection of information on relevant confounding variables is feasible. Second, the protective effects of different screening histories, such as interval since last smear and number of previous smears, can be evaluated. Finally, as described below, case-control data can provide information on the natural history of the disease process.

The first study of this kind compared the screening histories of 212 hospital cases of cancer with age-matched residential controls (Clarke & Anderson, 1979). A relative risk of 2.7 was observed in the unscreened women compared with those who had had at least one test in the preceding five years. In a subsequent study, Raymond *et al.* (1984) used as cases 186 registrations of invasive cancer in the Geneva Cancer Registry together with population controls; both groups had to have been resident in the canton for a period of 10 years, and information on screening history was obtained from records in the Cytology Centre. The degree of protection conferred by one negative smear was estimated to be 3.2. In Cali, Colombia (Aristizabal *et al.*, 1984), the case group comprised one-fifth of the incident invasive cancer cases notified to the cancer registry during 1977–1981, plus one-third as many prevalent cases under care in the radiotherapy department. Only 4.3% of these cases could recall having had a cervical smear during the 12–72 months before diagnosis, whereas 31% of age-matched neighbourhood controls reported a test in the same calendar period, implying a relative risk (case:control) of 9.9.

Evaluation of screening for breast cancer

In contrast to cervical cancer, the need for randomized trials to evaluate early detection of breast cancer was recognized before screening was widespread. Assessment of the effect is still based mainly on the Health Insurance Plan of New York study (Shapiro *et al.*, 1971). That trial was started in the early 1960s and involved the random allocation of 62 000 women aged 40–64 to a screened group, who were offered screening by physical examination and mammography in four successive years, or to a control group. The outcome measure upon which attention concentrated was breast cancer mortality rates in the two groups. The most recent results (Shapiro *et al.*, 1982) confirm a significant reduction in mortality 14 years after entry to the trial, principally in women aged over 50 at the time.

The Health Insurance Plan study was followed in the USA by the introduction of screening clinics in many locations; however, the results are frequently difficult to evaluate, as the population being screened is not defined and no 'control group' can be identified with which the results obtained in screened women can be compared. Furthermore, in many centres, no adequate follow-up procedures were installed. The Breast Cancer Demonstration and Detection Projects in the USA (Smart & Beahrs, 1979) provide a great deal of descriptive data on screened subjects, but their effectiveness in preventing late disease or mortality cannot be evaluated (Shapiro, 1978).

Results of improvements in the technical efficiency of mammography, using newer, lower-dose techniques, now need to be evaluated. Randomized studies evaluating mammography alone were initiated in Sweden in the mid-1970s (Andersson *et al.*, 1979; Tabar & Gad, 1981). The main endpoint is breast cancer mortality. A trial begun in Canada in the early 1980s (Miller *et al.*, 1981) will assess the effect of annual mammography plus physical examination in women aged 40–49, and the contribution of mammography to physical examination alone at age 50–59. The main endpoint of this study is comparison of mortality rates from breast cancer in the study groups with those in controls; the national cancer registry will be used as one source of follow-up data on the 90 000 subjects.

An example of a population-based study that was carefully planned so that, with reasonable assumptions, the results could be interpreted without a randomized control group is that of the DOM (Diagnostisch Onderzoek Mammacarcinoom) project in Utrecht (de Waard *et al.*, 1984) and a parallel study in Nijmegen (Hendriks, 1982). In Utrecht, a population-based cancer registry was established before the screening programme (four consecutive examinations by mammography and palpation) was introduced. Breast cancer incidence rates, by stage of disease, were determined for women in Utrecht aged 50–64 before and after introduction of screening, and rates in non-attenders compared with those in attenders. Further information is provided by the rate of interval cancers occurring in screened women in relation to the time between tests. The cancers found in the screened group (those detected at screening plus interval cases) were smaller in size and involved the axillary nodes less frequently than did those in the pre-screening and non-response groups. Initial evaluations of the effect on breast cancer mortality have been made by case-control studies in the two regions (Collette et al., 1984; Verbeek et al., 1984). Effects similar to, if not greater than, the effects seen in the Health Insurance Plan study were reported.

The trial of breast cancer screening in the United Kingdom (DHSS Working Group, 1981) involved health districts, rather than individuals, as the units of comparison. In two districts, a programme of annual physical examination and biennial mammography is offered; in two others, courses are given in breast self-examination; four other centres are used as controls. In one of the centres, randomization to screening or general publicity about breast self-examination alone is being carried out at the level of the general practitioner (Roberts et al., 1984). In this study, the national cancer registration scheme is being used to identify new cases of breast cancer, and the incidence rates act as an additional check on the significance of any change in mortality that may be observed.

Many studies have investigated the possible benefits of breast self-examination alone, some of which use cancer registries in order to have as unselected a case group as possible (Foster et al., 1978; Smith et al., 1980). Many of these report reduced size or favourable stage distribution of lesions detected by breast self-examination or in women who report practising breast self-examination (Foster et al., 1978; Greenwald et al., 1978; Huguley & Brown, 1981). It may also have a favourable impact on survival (Costanza & Foster, 1984). However, none of these studies had a control group, and the benefits due simply to lead time and length bias have to be guessed at.

Evaluation of other cancer screening procedures

Screening programmes have been introduced for other major cancers, most notably those of lung, large bowel and stomach, and some results are available to help interpret their effectiveness. For all of these tumour sites, screening aims to prevent deaths by detecting tumours in their early stages, at which time prognosis is more favourable. Their effectiveness must thus be judged by improvements in mortality, preferably in a controlled trial; data on incident cases (e.g., distribution, survival) must be interpreted with caution because of bias due to lead time and length bias, as already discussed.

For colon cancer, the effectiveness of screening by faecal occult blood testing is currently being investigated in two controlled trials (Gilbertson et al., 1980; Winawer et al., 1980). Both use mortality rates as endpoints but collect information on site, stage and outcome of incident cancers in subjects and controls.

Evaluation of screening for lung cancer by regular chest X-ray has been attempted in several randomized trials (Brett, 1968; Taylor et al., 1981). Although these demonstrate an apparent advantage in the stage and survival rates of cancers found in the screened group, there has been no effect on mortality or on the incidence of advanced disease. Lead time and length bias presumably account for this contrast.

Gastric cancer screening by photofluorography has been widely used in Japan. No randomized trial has been performed, and the evidence that some of the fall in stomach cancer mortality in Japan is attributable to screening is indirect, at best (Hirayama, 1978). In Miyagi prefecture, the population-based registry has been used to monitor incident cases and deaths from gastric cancer, and these have been linked to the data on screening (Yamagata et al., 1983). By defining a false positive result as an individual who is found to have gastric cancer within a year of screening, it was possible to evaluate the characteristics of the screening test (photofluorography); it had a sensitivity of 82.4%, but specificity was poor (77.2%), so that the predictive value of a positive examination is only 1.5%. Mortality from gastric cancer was lower in screened people than in those who had never been screened, but there was, of course, no attempt to randomize between the two groups. In a study in Osaka (Oshima et al., 1979), the cancer registry was used to follow up 33 000 screened people for incident cancers (those discovered at screening and interval cases) and deaths. The incidence, stage distribution and mortality rates in the screened population were compared with those from the general population of the prefecture. An excess of cases was seen in the screened group; this seems to have been due to detection of 'early' cases, since the number of advanced cancers was approximately that expected on the basis of population rates. Mortality from stomach cancer was reduced (observed:expected ratio, 0.91) in screened people, but so too was mortality from other, non-digestive-tract cancers (observed:expected ratio, 0.51), suggesting some strong selective factors in the screened population.

PLANNING SCREENING PROGRAMMES

The decision to implement a screening programme depends first of all upon a knowledge of its potential effectiveness, as discussed above. However, the efficiency of a programme (the ratio between outcome and cost) is an important practical consideration for health planners, and within screening programmes there are many variables that can be adjusted to influence the ratio of cost to outcome. These variables include, for example, the choice of screening test, follow-up procedures for different test results, choice of population to examine, frequency of examinations, and so on. Theoretically, intervention studies could be mounted to investigate the importance of changing such variables, but in practice they would have to be impossibly large, numerous or lengthy.

Models of screening programmes, either using a mathematical approach (Schwartz, 1978; Eddy, 1980) or computer simulation (Knox, 1973; Parkin, 1985), are a convenient approach to examining the possible effects of changing the parameters of a programme. All such models require data about the disease process and screening techniques; ideally, such data should be derived from observations in populations similar to that in which screening is proposed. The disease process is frequently modelled as a series of transition probabilities, and their derivation will require information on age-specific incidence rates of clinical cancer, and age-stage-specific survival curves (e.g., Schwartz, 1978; Parkin, 1985). These parameters can be derived only from population-based cancer registries. In addition, information must be obtained on the natural history of the disease in its preclinical, screen-detectable phase, and especially on the distribution of sojourn times in such phase(s) and the probability of regression to an earlier stage. If the average sojourn time in the preclinical detectable phase is long, then it is obvious that screening can take place at greater intervals than for diseases with short sojourn times, since there will be a greater probability that an individual who has entered the preclinical state will be detected by the screening test before this disease progresses to the clinical stage.

For cervical carcinoma, some estimates of average duration in the preclinical phase were made by comparing the age-specific prevalence of carcinoma *in situ* (in an unscreened population) with the age-specific incidence of invasive cancer. It was noted that there was an apparent age difference of some eight to ten years (Dunn & Martin, 1967; Fidler *et al.*, 1968). Since many in-situ lesions may not progress, and this proportion appears to be age-dependent, estimates obtained in this way are of dubious value.

The use of a population-based cancer registry, in association with an on-going screening programme, to identify interval cancers in a screened population has already been discussed. Walter and Day (1983) described how data on post-screening incidence can be used to determine two parameters of great importance in screening: the distribution of sojourn times and test sensitivity. The importance of sojourn time in determining frequency of testing has been mentioned; sensitivity of a test defines the proportion of diseased individuals who are missed on screening (Thorner & Remein, 1961), and is hence also an important parameter in design of programmes. Immediately after a negative screening result, the incidence of invasive cancer is low (equal to the initial incidence multiplied by the false-negative rate), and this risk will increase slowly as time elapses since the negative smear. Figure 4 illustrates how the post-screening incidence rate (incidence of interval cancers) increases with time, and shows the relationship between post-screening incidence and the two determinant factors: the distribution of sojourn times and the false-negative rate. Also shown is the effect of repeating screening tests, one result of which should be to reduce the number of false negatives.

Walter and Day (1983) have used data from the Health Insurance Plan study on incidence rates of interval cancer by time since previous examination and number of screening tests to estimate test sensitivity, and to examine possible lead-time distributions. They suggest that the sensitivity of mammography plus physical examination in this study was 82%, and that an exponential distribution of the detectable preclinical place with a mean of 1.7 years provides the best fit to the data. A similar analysis using incidence data is proposed in the DOM project in Utrecht (de Waard *et al.*, 1984).

For cervical cancer, Boyes *et al.* (1982) calculated false-negative error of cytology tests using an identical approach on the British Columbia data; false-negative error (excluding errors in interpretation of smears) was estimated to be about 10%. The IARC collaborative study (Miller *et al.*, 1985) is using data from this and other centres from which there are accurate and comprehensive cytology records and from population-based cancer registers to provide estimates of sojourn time and its distribution.

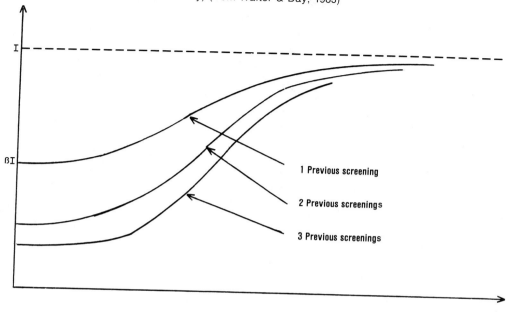

Fig. 4. Models of post-screening incidence. Theoretical incidence rate at different times after one or more negative screening results. I, incidence before screening; βI, false-negative rate of screening results (1−sensitivity) (from Walter & Day, 1983)

MONITORING SCREENING PROGRAMMES

The first part of this paper described the requirements for evaluation of screening programmes, to determine whether or not a particular procedure is likely to be effective. The relevant outome measures are a reduction in mortality (or morbidity from advanced disease) and, in the case of cervical cancer screening, a reduction in incidence of clinically invasive cancer.

These represent long-term endpoints, however, and those responsible for planning and monitoring health services usually require more immediate and continuous assessment of the performance of a programme. Process measures indicate the level of activity of a screening programme, and, although they have only an indirect link with outcome, they can be used for screening procedures of known effectiveness to compare one programme with another in order to indicate whether a poor programme could be improved or should be disbanded (Cole & Morrison, 1978). Process measures may include numbers of tests performed, the attendance rate of the at-risk population or the proportions of attenders examined more than once. The yield of screening tests (percentage found positive) is related to test sensitivity and prevalence of the disorder in the population screened. The performance of the screening test itself can be monitored by determining its predictive value—the proportion of positive tests that are true positives. A high predictive value means that the test has high specificity (few false positives), that a high-prevalence population is being screened, or both.

Cancer registries can provide data on cancer incidence, on distribution of incident cases by stage of disease, and on survival rates. For screening programmes for cervical cancer, these measures can, in theory, provide a basis for monitoring programme success, particularly if supplemented by case-control studies. Falling rates of incidence of clinical cancer should accompany a successful programme; but they may be difficult to interpret if the underlying risk of cervical cancer in the population is not stable. Registration rates of carcinoma of the cervix can be a help in monitoring screening, both in providing an index of possible success in preventing clinical cancer, and in estimating possible trends in disease in the absence of screening (Parkin et al., 1985). Increasing use of the nomenclature 'cervical intra-epithelial neoplasia' for preinvasive lesions of the cervix is likely to pose problems for cancer registries in producing meaningful time trends of in-situ registrations (Alderson et al., 1983). One problem in using registry data to follow trends in incidence in the presence of a screening programme is that preclinical disease (micro-invasive and occult invasive) is registered along with clinically diagnosed cases. Even though such cases would presumably have surfaced as clinical cancer after a delay of a few years, so that crude all-ages incidence is not affected, there is an apparent shift to younger age groups. The result is that age-specific or cohort-specific rates may suggest an increase in incidence, or a fall in incidence may be masked. It is suggested, therefore, that registries carefully record the stage of cervical tumours, including micro-invasive and occult invasive cancer, and, if possible, record those cases in which diagnosis was a result of screening.

The use of case-control studies to evaluate the efficacy of screening in preventing invasive cancer has been described, and this approach has clear potential as a method of monitoring a given programme periodically. Population-based cancer registries can provide an unbiased case series for such studies.

When screening programmes search for early invasive disease (e.g., breast cancer), registries have a role to play in recording incidence of disease, by stage. Although diagnosis at earlier stages does not guarantee that screening has been effective, as has been discussed, if there is no favourable shift in stage distribution, then it is certain that no benefit is resulting. If the fact that a case was diagnosed by screening is recorded, stage distribution of screen-detected as opposed to non-screen-detected cases can be compared.

Survival rates by stage of disease may be obtainable if the cancer registry can follow up a substantial proportion of cases (at least of the tumour for which screening is conducted). If screening is to be effective, survival should be related to stage at diagnosis, and screen-detected cases should have a better prognosis than symptomatic cases. However, because of self-selection and length bias, this is not evidence that benefit will actually result.

The use of population-based registries to identify interval cancers has been described; the ratio of the incidence of interval cases to that before screening offers a further measure of the success of a particular programme.

CONCLUSIONS

The most widespread cancer screening activity at present is the use of the periodic 'Pap smear' for the prevention of invasive carcinoma of the cervix. Population-based cancer registries have played a major part in providing the evidence that has established the effectiveness of such screening. In addition, they provide data useful for the planning and monitoring of on-going programmes. For the other screening methods which aim to detect pre-invasive conditions (e.g., oral cancer), cancer registries can play a similar role. When the objective of screening is to find invasive cancer at an early stage (e.g., breast cancer), changes in mortality rates are the appropriate measure of effectiveness; nevertheless, the data that registries provide on incidence, stage distribution and survival can provide indirect measures of the outcome of screening.

REFERENCES

Alderson, M., Correa, P., Ford, J., Jensen, O.M., Kupka, K., Miller, A.B., Muir, C.S., Nectoux, J., Waterhouse, J.A. & Young, J.L. (1983) Cervical intraepithelial neoplasia. *Lancet, i,* 1166

Andersson, I., Andrén, L., Hildell, J., Linell, F., Ljungqvist, U. & Pettersson, H. (1979) Breast cancer screening with mammography: a population-based, randomized trial with mammography as the only screening mode. *Radiology, 132,* 273–276

Aristizabal, N., Cuello, C., Correa, P., Collazos, T. & Haenszel, W. (1984) The impact of vaginal cytology on cervical cancer risk in Cali, Colombia. *Int. J. Cancer, 34,* 5–9

Boyes, D.A., Worth, A.J. & Anderson, G.H. (1981) Experience with cervical screening in British Columbia. *Gynecol. Oncol., 12,* S143–S155

Boyes, D.A., Morrison, B., Knox, E.G., Draper, G.J. & Miller, A.B. (1982) A cohort study of cervical screening in British Columbia. *Clin. invest. Med., 5,* 1–29

Breslow, N.E. & Day, N.E. (1980) *Statistical Methods in Cancer Research,* Vol. 1, *The Analysis of Case-control Studies (IARC Scientific Publications No. 32),* Lyon, International Agency for Research on Cancer

Brett, G.Z. (1968) The value of lung cancer detection by six-monthly chest radiographs. *Thorax, 23,* 414–420

Bross, I.J. (1967) Pertinency of an extraneous variable. *J. chronic Dis., 20,* 487–497

Canadian Task Force on Cervical Cancer Screening Programs (1982) *Cervical Cancer Screening Programs 1982,* Ottawa, Department of Health & Welfare

Christopherson, W.M., Lundin, F.E., Mendez, W.M. & Parker, J.E. (1976) Cervical cancer control. A study of morbidity and mortality trends over a twenty year period. *Cancer, 38,* 1357–1366

Clarke, E.A. & Anderson, T.W. (1979) Does screening by 'Pap' smears help to prevent cervical cancer? A case control study. *Lancet, ii,* 1–4

Cole, P. & Morrison, A.S. (1978) *Basic issues in cancer screening.* In: Miller, A.B., ed., *Screening in Cancer (UICC Technical Report Series Vol. 40),* Geneva, International Union Against Cancer, pp. 7–39

Collette, H.J.A., Day, N.E., Rombach, J.J. & de Waard, F. (1984) Evaluation of screening for breast cancer in a non-randomised study (DOM Project) by means of a case-control study. *Lancet, i,* 1224–1226

Costanza, M.C. & Foster, R.S. (1984) Relationship between breast self-examination and death from breast cancer by age-groups. *Cancer Detect. Prev., 7,* 103–108

Cramer, D.W. (1974) The role of cervical cytology in the declining morbidity and mortality from cervical cancer. *Cancer, 34,* 2018–2027

DHSS Working Group (1981) Trial of early detection of breast cancer: description of method. *Br. J. Cancer., 44,* 618–627

Dickinson, L., Mussey, M.E., Soule, E.H. & Kurland, L. (1972) Evaluation of the effectiveness of cytologic screening for cervical cancer. I. Incidence and mortality trends in relation to screening. *Mayo Clin. Proc., 47,* 534–544

Draper, G.J. & Cook, G.A. (1983) Changing patterns of cervical cancer rates. *Br. med. J., 287,* 510–512

Dunn, J.E. & Martin, P.L. (1967) Morphogenesis of cervical cancer: Findings from the San Diego County cytology registry. *Cancer, 20,* 1899–1906

Eddy, D.M. (1980) *Screening for Cancer; Theory, Analysis and Design,* Englewood Cliffs, NJ, Prentice Hall

Feinleib, M. & Zelen, M. (1969) Some pitfalls in the evaluation of screening programs. *Arch. environ. Health, 19,* 412–415

Fidler, H.K., Boyes, D.A. & Worth, A.J. (1968) Cervical cancer detection in British Columbia. *J. Obstet. Gynaec. Br. Commonw., 75,* 392–404

Foltz, A.M. & Kelsey, J. (1980) The annual pap test: a dubious policy success. *World Health Forum, 1,* 105–116

Foster, R.S., Lang, S.P., Costanza, M.C., Worden, J.K., Haines, C.R. & Yates, J.W. (1978) Breast self-examination practices and breast cancer stage. *New Engl. J. Med., 299,* 265–270

Fraser, B.J. & Boyes, D.A. (1976) *The benefit of screening for cancer of the cervix in British Columbia.* In: Bostrom, H., Larsson, T. & Ljungstedt, N., eds, *Health Control in Detection of Cancer (Skandia International Symposia),* Stockholm, Almqvist & Wiksell, pp. 215–225

Gibertson, V.A., Church, T.R., Grewe, F.J., Mandel, J.S., McHugh, R.B., Schuman, R.M. & Williams, S.E. (1980) The design of a study to assess occult blood screening for colon cancer. *J. chronic Dis., 33,* 107–114

Greenwald, P., Nasca, P.C., Lawrence, C.E., Horton, J., McGarrah, R.P., Gabriele, T. & Carlton, K. (1978) Estimated effect of breast self-examination and routine physician examinations on breast cancer mortality. *New Engl. J. Med., 299,* 271–273

Guzick, D.S. (1978) Efficacy of screening for cervical cancer: a review. *Am. J. public Health, 68,* 125–134

Hakama, M. (1982) *Trends in the incidence of cervical cancer in the Nordic countries.* In: Magnus, K., ed., *Trends in Cancer Incidence,* New York, Hemisphere Press, pp. 279–292

Hendriks, J.H.C.L. (1982) *Population Screening for Breast Cancer by Means of Mammography in Nijmegen 1975–1980,* PhD Thesis, University of Nijmegen

Hirayama, T. (1978) *Outline of stomach cancer screening in Japan.* In: Miller, A.B., ed., *Screening in Cancer (UICC Technical Report Series No. 40),* Geneva, International Union Against Cancer, pp. 64–78

Huguley, C.M. & Brown, R.L. (1981) The value of breast self-examination. *Cancer, 47,* 989–995

Hutchinson, G.B. & Shapiro, S. (1968) Lead time gained by diagnostic screening for breast cancer. *J. natl Cancer Inst., 41,* 665–681

Johannesson, G., Geirsson, G., Day, N.E. & Tulinius, H. (1982) Screening for cancer of the uterine cervix. The effect of mass screening in Iceland, 1965–78. *Acta obstet. gynecol. scand., 61,* 199–203

Knox, E.G. (1973) *A simulation system for screening procedures.* In: McLachlan, G., ed., *The Future—and Present Indicatives (Problems and Progress in Medical Care, Series IX),* Oxford, Nuffield Provincial Hospitals Trust, pp. 19–55

Knox, E.G., ed. (1979) *Epidemiology in Health Care Planning,* London, Oxford Medical Publications

Laskey, P.W., Meigs, J.W. & Flannery, J.T. (1976) Uterine cervical carcinoma in Connecticut, 1953–1973: Evidence for two classes of invasive disease. *J. natl Cancer Inst., 57,* 1037–1043

Lynge, E. (1983) Regional trends in incidence of cervical cancer in Denmark in relation to local smear-taking activity. *Int. J. Epidemiol., 12,* 405–413

Miller, A.B. (1982) *The Canadian experience of cervical cancer: Incidence trends and a planned natural history investigation.* In: Magnus, K., ed., *Trends in Cancer Incidence,* Washington DC, Hemisphere Publications, pp. 311–320

Miller, A.B., Lindsay, J. & Hill, G.B. (1976) Mortality from cancer of the uterus in Canada and its relationship to screening for cancer of the cervix. *Int. J. Cancer, 17,* 602–612

Miller, A.B., Howe, G.R. & Wall, C. (1981) The national study of breast cancer screening. Protocol for a Canadian randomized controlled trial of screening for breast cancer in women. *Clin. invest. Med., 4,* 227–258

Miller, A.B., Hakama, M. & Day, N.E. (1985) *Screening for Cancer of the Cervix (IARC Scientific Publications),* Lyon, International Agency for Research on Cancer (in press)

Morrison, A.S. (1982) Case definition in case-control studies of the efficacy of screening. *Am. J. Epidemiol., 115,* 6–8

Oshima, A., Hanai, A. & Fujimoto, I. (1979) Evaluation of a mass screening programme for stomach cancer. *Natl Cancer Inst. Monogr., 53,* 181–186

Parkin, D.M. (1985) A computer simulation model for the practical planning of cervical cancer screening programmes. *Br. J. Cancer, 51,* 551–568

Parkin, D.M., Collins, W. & Clayden, A.D. (1981) Cervical cytology screening in two Yorkshire areas: pattern of service. *Publ. Health (London), 95,* 311–321

Parkin, D.M., Nguyen-Dinh, X. & Day, N.E. (1985) The impact of screening on the incidence of cervix cancer in England and Wales. *Br. J. Obst. Gynaecol., 92,* 150–157

Pedersen, E., Hoeg, K. & Kolstad, P. (1971) Mass screening for cancer of the uterine cervix in Ostfold county, Norway: An experiment. *Acta obstet. gynecol. scand.,* Suppl. 11, 1–18

Raymond, L., Obradovic, M. & Riotton, G. (1984) Une étude cas-témoins pour l'évaluation du dépistage cytologique du cancer du col utérin. *Rev. fr. Epidémiol. Santé publ., 32,* 10–15

Roberts, A. (1982) Cervical cytology in England and Wales 1965–1980. *Health Trends, 14,* 41–43

Roberts, M.M., Alexander, F.E., Anderson, T.J., Forrest, A.P.M., Hepburn, W., Huggins, A., Kirkpatrick, A.E., Lamb, J.W., Lutz, W. & Muir, B.B. (1984) The Edinburgh randomised trial of screening for breast cancer: Description of method. *Br. J. Cancer, 50,* 1–6

Schwartz, M. (1978) An analysis of the benefits of serial screening for breast cancer based upon a mathematical model of the disease. *Cancer, 41,* 1550–1564

Shapiro, S. (1978) *Efficacy of breast cancer screening.* In: Miller, A.B., ed., *Screening for Cancer (UICC Technical Report Series Vol. 40),* Geneva, International Union Against Cancer, pp. 133–157

Shapiro, S., Strax, P. & Venet, L. (1971) Periodic breast cancer screening in reducing mortality from breast cancer. *J. Am. med. Assoc., 215,* 1777–1785

Shapiro, S., Venet, W., Strax, P., Venet, L. & Roeser, R. (1982) Ten to fourteen-year effect of screening on breast cancer mortality. *J. natl Cancer Inst., 69,* 349–355

Smart, C.R. & Beahrs, O.H. (1979) Breast cancer screening results as viewed by the clinician. *Cancer, 43,* 851–856

Smith, E.M., Francis, A.M. & Polissar, L. (1980) The effect of breast self-exam practices and physician examinations on extent of disease at diagnosis. *Prev. Med., 9,* 409–417

Tabar, L. & Gad, A. (1981) Screening for breast cancer: The Swedish trial. *Radiology, 138,* 219–222

Taylor, W.F., Fontana, R.S., Uhlenhopp, M.A. & Davis, C.S. (1981) Some results of screening for early lung cancer. *Cancer, 47,* 1114–1120

Thorner, R.M. & Remein, Q.R. (1961) Principles and procedures in the evaluation of screening for disease. *Public Health Monogr., 67*

Verbeek, A.L.M., Hendriks, J.H.C.L., Holland, R., Mravunak, M., Sturmans, F. & Day, N.E. (1984) Reduction of breast cancer mortality through mass screening with modern mammography. *Lancet, i,* 1222–1224

de Waard, F., Collette, H.J.A., Rombach, J.J., Baanders van Halewijn, E.A. & Honing, C. (1984) The DOM project for the early detection of breast cancer, Utrecht, The Netherlands. *J. chronic Dis., 37,* 1–44

Walter, S.D. & Day, N.E. (1983) Estimation of the duration of a preclinical state using screening data. *Am. J. Epidemiol., 118,* 865–886

Walton Report (1976) Cervical cancer screening programs. II. Screening for carcinoma of the cervix. *Can. med. Assoc. J., 114,* 1013–1026

Weiss, N.S. (1983) Control definition in case control studies of the efficacy of screening and diagnostic testing. *Am. J. Epidemiol., 118,* 457–460

Wilson, J.M.G. & Junger, G. (1968) *Principles and Practice of Screening for Disease (Public Health Papers No. 34),* Geneva, World Health Organization

Winawer, S.J., Andrewes, M., Flehinger, B., Sherlock, P., Schottenfeld, D. & Miller, D.G. (1980) Progress report on controlled trial of fecal occult blood testing for the detection of colorectal neoplasia. *Cancer*, **43**, 2959–2964

World Health Organization (1969) *Early Detection of Cancer (Technical Report Series No. 422)*, Geneva

Yamagata, S., Sugawara, N. & Hisamichi, S. (1983) *Mass screening for cancer in Japan—Present and future*. In: Yamagata, S., Hirayama, T. & Hisamichi, S., eds, *Recent Advances in Cancer Control (Proceedings of the 6th Asia Pacific Cancer Conference, Sendai, Japan, Sept. 27–30, 1983)*, Amsterdam, Excerpta Medica, pp. 33–45

5. THE CANCER REGISTRY AS A TOOL FOR DETECTING INDUSTRIAL RISKS

O.M. JENSEN

Danish Cancer Registry, Institute of Cancer Epidemiology, Danish Cancer Society, Landskronagade 66, DK-2100 Copenhagen, Denmark

SUMMARY

The cancer registry plays a role both in the generation and testing of hypotheses of cancer risks in relation to occupation. Most industrial risks have been detected by alert clinicians; but an increased risk of nasal cancer in relation to wood work was also observed by the Danish Cancer Registry. Although 55 out of 81 cancer registries in the world record occupation, this information is not fully exploited. Emerging associations should be compared internationally and followed by independent testing in studies of the cohort as case-control studies. Cancer registries may be useful in both, but to that end records must be stored for decades both in industry and in the registry, and they must carry personal identification. Linkage of records for medical purposes should be legal and subject to supervision by a data protection board.

INTRODUCTION

In accordance with previous estimates by Higginson and Muir (1979), it has been suggested by Doll and Peto (1981) that some 2–8% of all cancers today in economically developed countries are caused by occupational exposures (Table 1). It is not the purpose of this paper to discuss at length whether this proportion—if correct—justifies a heavy investment of human and financial resources into research on occupational cancer. It is worth pointing out, however, that (1) there is a widespread demand in society that occupational activities not include hazards to human health, and (2) cancer prevention in the occupational field may largely be accomplished through regulatory action (e.g., change of products, modification of

Table 1. Proportions of all cancer deaths attributable to various factors[a]

Factor or class of factors	Percent of all cancer deaths (acceptable estimates)
Tobacco	25 –40
Alcohol	2 – 4
Diet	10 –70
Reproductive and sexual behaviour	1 –13
Occupation	2 – 8
Geophysical factors	2 – 4
Medicines and medical procedures	0.5– 3

[a] From Doll & Peto (1981)

processes and improvements in hygienic standards). Both of these aspects are important prerequisites for the success of public health actions; this clearly distinguishes intervention in the field of occupational cancer from, for instance, modification of smoking or dietary habits.

In addition to this first and direct objective of research into occupational cancer, results from heavily exposed groups may indicate risks incurred by the general population, which is often exposed to much smaller doses. Otherwise undetectable risks may prove numerically important when sufficiently large populations are exposed.

Finally, an increasing number of investigations are carried out on the carcinogenic potential of single chemical compounds and physical agents in the occupational environment, in both short- and long-term experimental systems. Both cancer research and society, however, have an interest in human studies, whenever they are feasible, to shed light on the magnitude of risk experienced by humans at the dose level to which they are exposed.

Both in relation to other areas of research in humans and to other fields of cancer research, there is thus a clear interest in the study—qualitative and quantitative—of cancer risks in relation to occupational exposures. The role of the cancer registry is four-fold:

(1) The registry may serve as a basis for the identification of high- (and low-) risk occupational groups and thus be useful for the generation of etiological hypotheses.

(2) The registry may serve as a data source for cohort (follow-up) and case-control studies to test etiological hypotheses.

(3) The registry may provide reference material for cohort and case-control studies.

(4) The cancer registry is an institute for cancer epidemiology. It should have epidemiological and statistical expertise for the planning, execution, analysis and interpretation of epidemiological studies, including those of occupation.

In the following, the use of geographical studies for the identification of industrial hazards is not addressed. That approach has in some instances proved valuable (Hoover & Fraumeni, 1975; Waterhouse, 1975), but the direct study of occupational groups is more likely to give a readily interpretable result.

IDENTIFICATION OF HIGH-RISK OCCUPATIONS—
HYPOTHESIS GENERATION

Studies without denominators

The history of occupational cancer shows how immensely important is the observation by an alert clinician of the clustering of cases in single occupations. One of the earliest examples comes from Germany with the description by Paracelsus in 1531 among miners in Schneeberg and St Joachimstal of *mala metallorum*—a pulmonary disease later identified as lung cancer and related to the mining of radioactive ore (Clemmesen, 1965). The demonstration of chimney sweeps' scrotal cancer (Pott, 1775) and of bladder cancer in aniline dye workers (Rehn, 1895) date back to the end of the eighteenth and the end of the nineteenth centuries, respectively. Nasal adenocarcinomas in wood workers (MacBeth, 1965), asbestos-related pleural mesothelioma (Merewether, 1949) and vinyl chloride as an underlying cause of haemangiosarcoma of the liver (Creech & Johnson, 1974) are more recent examples.

Neither cancer registries nor other medical information systems played a role when these risks were initially observed. However, MacBeth's (1965) observation of nasal tumours among wood workers gained credibility from its immediate confirmation by the Oxfordshire Cancer Registry (Acheson *et al.*, 1968). It is a common characteristic of these observations that they combine a high risk of a rare cancer in a very specific occupation.

Could these observations have been made by a cancer registry? This question has been evaluated with regard to nasal cancer using information in the Danish Cancer Registry, as it is routinely reported by clinical departments.

The classification of occupation was introduced in 1943, categorizing patients by occupation at the time of cancer diagnosis. In 1980, it was found that the categories were too broad for the derivation of useful information. Subgroups not previously coded as such were therefore given individual code numbers, retaining in a hierarchial system the overall categorization used since 1943. In 1980, the ICD-O tumour classification (World Health Organization, 1976) was introduced, in which each tumour is classified by topography, morphology and behaviour. This new classification system was applied in retrospect to incidence data from 1978 onwards (Danish Cancer Registry, 1982).

For the three-year period 1978–1980, the risk of nasal cancer was evaluated in relation to occupations involving exposure to wood dust. The analysis is based entirely on cancer registry information, as no population denominators comparable to the Cancer Registry classification are available. It is apparent from Table 2 that among 89 men with a specified occupation and a malignant tumour of the nasal cavities, seven or 7.9% were wood workers. This compares with 3.7% of wood workers among 3662 colorectal cancer patients, and it yields a relative risk of 2.2 for the development of nasal cancers. There were 14 adenocarcinomas, with a relative risk of 7.1. Two of these occurred in furniture makers, corresponding to 14.3% of the nasal adenocarcinomas, with a risk 640 times higher than the risk in other occupations. Although the numbers are small, and although the study was carried out

Table 2. Cancer of the nasal cavities 1978–1980 and occupational exposure to wood dust in males in Denmark

Site of tumour	No.	Wood workers		Furniture makers	
		%	Relative risk	%	Relative risk
Controls[a]	3662	3.7	1.0	0.03	1.0
Nasal cancer, all	89	7.9	2.2[b]	2.2	(86.0)[c]
Nasal cancer, adenocarcinoma	14	21.4	7.1[b]	14.3	(641.1)[c]

[a] People with cancer of the colon and rectum
[b] Stratified for age; $p < 0.05$
[c] Risk estimates based on small numbers

after demonstration of the association and the risk estimates are unstable, three years of data were sufficient to indicate an unusual situation.

Considering that 55 out of 81 registries that reported data to the fourth volume of *Cancer Incidence in Five Continents* (Waterhouse *et al.*, 1982) recorded occupation in some form, it is tempting to suggest that so far too little use has been made of this information. In the Danish Cancer Registry, occupational information has been recorded every day for 40 years, and no or little use has been made of the information now available for some 700 000 tumours. In the absence of population denominators, proportional analysis should not be overlooked as a quick and inexpensive method for cancer registries to screen available data. It may be supplemented by derivation of relative risks by case-control analysis.

Studies with denominators

The use of cancer registry data for the calculation of incidence rates for occupational groups does not differ from the use of information from death certificates. Experience with the latter is much more widespread than the use of cancer registry information for this purpose. Occupation as given on death certificates in England and Wales has been related to the number of persons at risk known from the census data (Office of Population Censuses and Surveys, 1978). There are well-known difficulties related to the possible lack of comparability between numerators and denominators, due, for instance, to the tendency to notify a higher social status after death. Although few—if any—associations between occupational exposure and cancer risk have first been suggested on the basis of this data source, this type of information is useful for checking and monitoring trends in known risks.

Some of the possible biases in the type of material represented by the occupational mortality experience in England and Wales have been avoided in Scandinavia, where information from population censuses in Denmark, Finland, Norway and Sweden has been linked with cancer registry material as well as with mortality data. The Swedish material is derived from the so-called Environment-Cancer Registry, from which a number of previous as well as new associations have emerged (National Board of Health and Welfare, 1980; Socialstyrelsen, 1980).

Jensen, O. M. (1979) Cancer morbidity and causes of death among Danish brewery workers. *Int. J. Cancer, 23,* 454–463

Jensen, O. M. (1980) Cancer risk from formaldehyde (Letter). *Lancet, ii,* 480–481

Jensen, O. M. & Krüger Andersen, S. (1982) Lung cancer risk from formaldehyde (Letter). *Lancet, i,* 913

Jensen, O. M., Knudsen, J. B., Sørensen, B. L. & Clemmesen, J. (1985) Artificial sweeteners and absence of bladder cancer risk in Copenhagen. *Int. J. Cancer* (in press)

Lynge, E., Andersen, O. & Kristensen, T. S. (1983) Lung cancer in Danish butchers (Letter). *Lancet, i,* 527–528

MacBeth, R. (1965) Malignant disease of the paranasal sinuses. *J. Laryngol., 79,* 592–612

Magnus, K., Andersen, A. & Høgetveit, A. C. (1982) Cancer of respiratory organs among workers at a nickel refinery in Norway. *Int. J. Cancer, 30,* 681–685

Medical Research Council (1982) *Job Exposure Matrices. Proceedings of a Conference held in April 1982 at the University of Southampton, Southampton*

Merewether, E. R. A. (1949) *Asbestosis and carcinoma of the lung.* In: *Annual Report of the Chief Inspector of Factories for the Year 1947,* London, Her Majesty's Stationery Office, pp. 79–81

National Board of Health and Welfare (1980) *The Swedish Cancer Environment Registry, 1961–1977,* Stockholm, Committee for the Cancer Environment Registry

Office of Population Censuses and Surveys (1978) *Occupational Mortality. The Registrar General's Decennial Supplement for England and Wales, 1970–72,* London, Her Majesty's Stationery Office

Pedersen, E., Høgetveit, A. C. & Andersen, A. (1973) Cancer of respiratory organs among workers at a nickel refinery in Norway. *Int. J. Cancer, 12,* 32–41

Percy, C., Stanek, E. & Gloeckler, L. (1981) Accuracy of cancer death certificates and its effect on cancer mortality statistics. *Am. J. public Health, 71,* 242–250

Pott, P. (1775) *Chirurgical Observations related to the Cataract, the Polypus of the Nose, the Cancer of the Scrotum, the Different Kind of Ruptures,* London

Rehn, L. (1895) Blasen-Geschwülste bei Fuchsin-Arbeitern. *Arch. klin. Chir., 50,* 588–600

Socialstyrelsen (1980) *Cancer Risk och Miljöfaktorer (Rapport til Arbetarskyddsfonden),* Stockholm

Storm, H. H., Schou, G. & Møller, C. D. (1982) *Validity of Death Certificates and Effects on Mortality Statistics in Denmark in Relation to Cancer Diagnoses. IACR Biennial Meeting, Seattle, September 1982*

Teppo, L., Hakama, M., Hakulinen, T., Lehtonen, M. & Saxén, E. (1975) Cancer in Finland 1953–1970: Incidence, mortality, prevalence. *Acta pathol. scand.,* Section A, Suppl. 252

Waterhouse, J. A. (1975) *The use of cancer registry data in the investigation of industrial carcinogenic hazards.* In: Grundman, E. & Pedersen, E., eds, *Cancer Registry,* Berlin, Springer, pp. 148–154

Waterhouse, J., Muir, C. S., Shanmugaratnam, K. & Powell, J., eds (1982) *Cancer Incidence in Five Continents Vol. IV (IARC Scientific Publications No. 42),* Lyon, International Agency for Research on Cancer

World Health Organization (1976) *International Classification of Diseases for Oncology. First Edition,* Geneva

6. PLANNING SERVICES FOR THE CANCER PATIENT

R.J. WRIGHTON
Mid-Downs Health Authority, Haywards Heath, West Sussex, UK

This chapter discusses the information requirements for planning services for cancer patients, with particular reference to the use of cancer registration data.

PLANNING HEALTH SERVICES

Most human activities, if they are to achieve their desired objective without waste of resources, require planning. This applies to relatively simple tasks carried out by the individual, to the most complex tasks of industry or government. The planning process requires decisions to be made in the present that will lead to desired effects in the future, enabling the objective to be reached. The planning process is particularly important in major state undertakings, such as the provision of health care, where increasing pressures on national resources allow little or no expansion to meet rising demand.

A good information base is of fundamental importance to rational and effective planning. Accurate information is required about the situation existing at the start of planning; forward estimates must be available of the changes likely to occur in major relevant factors over the timescale that is being planned for; information is also needed for monitoring the effects of implementation of the plan.

In the planning of health care services for any group, accurate information on morbidity and mortality is essential. Cancer patients are one of the few care groups for whom good morbidity data can be made available, through the cancer registration procedure.

The planning process

Assuming the existence of an agreed policy and stated objectives, planning may take place on a number of different timescales and at different levels within the organization.

Timescales:
Short-term (tactical) planning, or programming, usually covers a period up to a maximum of three years ahead, is mainly concerned with meeting current needs and takes place within an already largely determined pattern of resource availability.

Medium-term (strategic) planning covers periods up to ten years ahead, allows scope for recognizing new needs, re-allocating resources or obtaining new resources.
Long-term (perspective) planning covers periods up to twenty years ahead and is concerned with broad measures related to overall policy objectives.

Levels of planning:
Local—an individual community or local government area.
District—covering more than one community.
Regional—covering more than one district.
National—central government or national agency.
International—more often advisory to national governments than executive.

In theory, planning on any timescale can be carried out at any level, but in practice the lower the level at which planning is done the shorter the timescale on which it is possible to plan. Perspective planning and some strategic planning, because of its implications for shifting resources, has to be carried out at a national level; much strategic planning is carried out at a regional level, within the confines of national plans; district or local planning is largely concerned with tactical planning within the confines of regional plans. In any planning system based on this type of tiered structure, there must clearly be mechanisms for consultation between the different levels. The policy on which the national plans are based must be clearly understood at the local level, and national agencies must be aware of how their policy is being interpreted by local planners.

Steps in the planning process

At whatever level planning is carried out, the stages of the process are essentially the same. In a fully structured planning system, the sequence of stages is repeated at intervals, usually not more frequently than annually, so that a repeating cycle of work is established and the planning process becomes continuous.

The existing situation must first be described *(situation analysis)* and the problems identified *(problem definition)*. The defined problems are ranked according to the perceived urgency and feasibility of solution *(setting priorities)*, and relevant achievement targets are identified *(setting objectives)*. The various ways of reaching the targets are considered *(designing strategies)* and the most appropriate one identified *(analysis of alternative strategies)*. The necessary finance, personnel and equipment to implement the chosen strategy are allocated *(resource allocation)*, and the plan is *implemented*. The implementation is *monitored* and the plan adjusted as necessary. The effects of the plan are *evaluated* by means of a mechanism defined before the plan is implemented, and the results are fed into a subsequent planning cycle.

Fully structured planning systems have been introduced into the health care field only in recent years, and, as yet, very few countries use such systems to plan health care as a whole or to plan for specific care groups.

Planning cancer services

Services for cancer patients are planned according to the general principles of health care planning outlined above. The mechanisms employed will vary in detail from country to country, according to the structure of the health services and the national and regional planning systems. It is likely that there will be a national policy-forming body, and elements of the service may also be planned at the national level. In some countries, particularly in eastern Europe, a considerable amount of medium-term planning takes place at the national level, but more commonly this is a regional or even local function. There are also marked differences depending on whether the services for cancer patients are regarded as an integral part of the general health services and are planned accordingly or are seen as a distinct service and planned separately from the general services.

In England and Wales, services for cancer patients are planned within the context of general health care provision. Policy guidelines specific to cancer services are promulgated by central government on the advice of a national cancer advisory committee; specialist advisory machinery also exists at the regional level. Except for certain specific services, notably radiotherapy, planning at the regional and district level for the care of cancer patients is contained within the acute sector, and cancer patients are not identified as a specific care group.

INFORMATION REQUIREMENTS FOR HEALTH CARE PLANNING

The situation analysis and problem definition steps in planning depend on the availability of good cross-sectional information about the current situation and longitudinal information from which to identify trends and on which projections may be based. Good information is similarly required for monitoring and evaluation. The types of information required will vary with the sector for which planning is being done, but certain requirements are common.

Information on the population served is obviously of fundamental importance, and data on the structure of the population in relation to age, sex and race or country of origin is required. Fertility and mortality data are needed for population projection estimates. The geographical distribution of the population and its dynamics in terms of health care referral patterns need to be known, and a data set similar to the national one should also be available at regional and subregional levels.

Information on disease incidence by pathological type, sex and age should ideally be available. If the population mortality data are similarly subdivided by pathological cause of death, the relationship between incidence and mortality (sometimes termed the 'lethality' of a particular disease) enables survival estimates to be derived, and from a series of annual figures incidence, mortality and survival projections may be made. The geographical distribution of incidence and mortality within the population to be planned for may be useful in some instances. In practice, accurate data on disease incidence are available for only a limited number of conditions, and reliance may have to be placed on estimates based on sample surveys. For some diseases, it may be possible to use mortality data as a basis for an estimation of disease load.

This information on the population and its disease experience must be supplemented by information on the resources available for health care and on the relationship between the health care services and other relevant services funded from the same or other sources (e.g., social services, voluntary agencies).

Information requirements for evaluation

The information needed to determine whether plans have been successful in achieving their aims will vary according to the timescale and level of planning. The ultimate objective of health care planning is the reduction of morbidity and mortality from disease. The assessment of efficacy therefore requires information on changes in the indices of disease incidence, mortality, duration and (quality of) survival subsequent to the implementation of the plans, supplemented by information on resources used.

In practice, however, evaluation of health services cannot rest on these criteria alone. Changes in incidence, mortality and survival from any disease may take a very long time to become measurable, and countless other factors may influence these indices, apart from the health services. Other, short-term measures have to be introduced to measure the effect of change in health service provision. A variety of such measures exist. In general they measure activity (or 'process')—for example the number of cases treated, or percentage of the target population informed—rather than outcome, and the link between the two may be somewhat indirect. The development of more generally applicable performance indicators is at an early stage but, if successful, will make the overall evaluation of health service planning initiatives somewhat easier.

Sources of information

Demographic data are collected on a regular basis in most developed countries, mainly through the mechanism of the quinquennial population census. Changes in the indigenous population are monitored continuously through statutory birth and death registration systems. In some countries, information on the cause of death obtained from the death certification procedure can be linked with information on preceding life events, such as cancer registration, enabling survival rates to be calculated on a national basis.

Disease incidence data covering the whole population are available only for those diseases for which special arrangements have been made for the notification of the disease. This applies to certain infectious diseases for which, because of public health hazards, notification is statutorily required (e.g., tuberculosis); to certain rare diseases for which, because of their small numbers and easily identifiable nature, information on incidence is fairly readily available (e.g., phenylketonuria); and, in many countries, to cancer (see review by Weddel, 1973). Where national registration schemes exist or where there are good regional schemes, cancer registration provides a disease incidence base for planning. As a result of such schemes, there are better disease incidence data for cancer than for any other major disease group. Incidence data on other major disease groups are available only from sample surveys, which may be national or regional, continuous or intermittent.

Resource information will normally be available from management information systems. It will cover the finance, manpower, buildings and equipment available and used for the provision of health care. It may not be possible routinely to relate resource information to a specific disease group, and for this special studies may be required, for example, on the costing of cancer chemotherapy (Wrighton, 1979) and bone-marrow transplantation (Kay et al., 1980).

USE OF CANCER REGISTRATION DATA IN SERVICE PLANNING

The population-based cancer registry has long been acknowledged as a useful source of information for service planning purposes (Pedersen, 1962; Office of Population Censuses and Surveys, 1981; Donnan, 1982), but few examples of such usage have been published. In the remainder of this chapter the use of cancer registration data in service planning will be illustrated with reference to the cancer registration scheme in England and Wales and the planning of cancer services of the National Health Service.

Cancer registration in England and Wales

Information on the incidence of cancer began to be collected in England in 1923, and from 1945 a national cancer registration scheme has been in operation. England and Wales have one of the few population-based cancer registration schemes with national coverage in the world. The scheme is based on fifteen regional registries, fourteen in English National Health Service regions and one in Wales. The mechanisms for data retrieval at the hospital and primary care levels differ between the regions, but in all cases the data are collated at the regional level and the verified data are transmitted to the National Cancer Registry, which is held at the Office of Population Censuses and Surveys. The accuracy and completeness of coverage are known to vary between the regions. The level of ascertainment in the majority of regions is 90% or higher, but in one or two regions it may be as low as 70% (Office of Population Censuses and Surveys, 1981). Nevertheless, the scheme provides a most valuable source of information on the incidence of the disease.

The basic data provide the numbers of cases of cancer occurring annually in the population by site, sex and age. The fact of registration is recorded on the patient's medical record at the National Health Service Central Registry, which provides automatic linkage with death notification and allows cause of death and duration of survival to be known for virtually all patients registered. Overall survival data by site of disease are therefore available for the total population. Furthermore, linkage of registration information with death certification provides a safety net to ensure that survival data on patients lost to clinical follow-up, for whatever reason, barring emigration, are retrievable.

By definition, the information is available at both national and regional levels and can therefore be used in the development of national policy and plans or of more detailed regional plans. Registration data can be broken down into smaller population units for particular purposes, but the value of this exercise in relation to planning is doubtful. Variations in incidence by local area provide useful pointers for further research into etiological factors; there are also interesting regional differences in survival rates (Silman & Evans, 1981), although considerable caution is necessary in their interpretation (see Hanai & Fujimoto, this volume).

The national scheme calls for a fairly limited data set on each patient registered, but some regional registries collect more detailed information about each patient from their own populations—for example, on stage at presentation or treatment given. This information can be of great value in the monitoring and evaluation of services at the regional level, and in those regions with large populations the data may be extrapolated, with care, to the national population.

Policy and planning at national level

The availability of good disease incidence data is essential to the development of national policy and allows strategic and perspective planning of services on a more rational basis than is possible when such information is not available and reliance has to be placed on population and mortality data with estimates only of disease incidence. Gandy (1980) discusses the use of the standardized mortality ratio (SMR) as a proxy measure of cancer morbidity for resource allocation purposes. He concludes that for neoplasms with short survival time, the SMR is a reasonable approximation, but where survival times are long the standardized registration ratio is a more appropriate measure of cancer morbidity. This point is also discussed by Enterline *et al.* (1982). For the purpose of interregional resource allocation, the use of cancer registration data requires uniformity in completeness of registration between regions.

Considerable emphasis was placed on the need for a comprehensive cancer registration scheme during a recent review of policy for cancer services in England and Wales by the national advisory committee (the Standing Sub-Committee on Cancer, 1984). In the review, registration data were used to define the 'size of the problem', and trends in incidence and survival were used to make projections of future needs. By relating age-specific incidence and survival data to projected changes in population age structure, Hakama (1980) predicted major increases in cancer incidence, and hence service need, by the end of the century.

Planning radiotherapy services

Radiotherapy is a fundamental requirement for cancer care and must be available to all patients who need it. Radiotherapy equipment is expensive, and, together with the special building requirements, setting up a department and maintaining it at optimal efficiency becomes a costly exercise. Therefore, the minimum number of departments compatible with a reasonably accessible service must be set up, and all should be used to their full capacity.

From the cancer registration scheme, it is known that the overall, all-age national incidence rate in England and Wales is approximately 4000 per million population;

Table 1. Radiotherapy capacity requirements[a]

Annual incidence rate	4 000 per million population
50% require radiotherapy	2 000 per million population
Maximum capacity of linear accelerator	800 courses of treatment/year
Maximum capacity of cobalt unit	400 courses of treatment/year

Given that: no radiotherapy department should be equipped with less than two machines;

a linear accelerator requires linear accelerator backup

National guideline: radiotherapy departments should serve catchment populations of not less than 1 million

[a] From Department of Health and Social Security (1978)

and from information on the type of treatment given to cancer patients, also collected from the cancer registration scheme, it is known that approximately half of those patients will require radiotherapy at some stage in their illness. Information is available from radiotherapy workload studies on the capacity of a linear accelerator or cobalt unit under normal working conditions. Using these two sources of data, the radiotherapy capacity required for an average population of one million can be calculated, and this figure may then be used as a general guideline from which to develop a national policy for radiotherapy services (Table 1).

Bearing in mind the national guideline, regional policies can then be developed using regional incidence data which can be compared with national data and adjustments made as necessary. More detailed breakdowns of regional data by age and sex and by histological type may make possible more precise calculation of need (Cohen et al., 1982). In England and Wales, the degree of variation in incidence patterns between regions is small and unlikely materially to influence radiotherapy requirements, but this may not be so in countries with more diverse populations.

The availability of the disease incidence base makes it possible to calculate service needs in a much more precise way than would be possible without it. The distribution of radiotherapy services in Greater London was the subject of a major study by the London Health Planning Consortium in 1979. As a technologically based service, much of the early development of radiotherapy took place in the teaching hospitals, and radiotherapy departments have been maintained there as an integral part of the teaching function. However, there is a heavy concentration of teaching hospitals in London, and 11 of the undergraduate and two of the postgraduate teaching hospitals within a seven-kilometre radius of Oxford Circus have radiotherapy departments. The resident population within that area is 1.2 million and projected to fall; the catchment population is considerably greater, about 7 million, but that population also has access to a further four radiotherapy departments outside central London. The study concluded that there was considerable over-provision in terms of departments and that some of the departments were consistently under-used.

Having defined the problem, the study group was expected to make recommendations on the rationalization of the service; but good information on need is essential to that task. London is covered by three different regional cancer registries, and the study group found that, while there were good data on cancer incidence patterns, treatment and outcome from one registry, the data available from the others was incomplete and out of date and could not be used in a detailed analysis of needs. Any possible variation from national averages that the population of Greater London might show and which might influence their needs for radiotherapy services could not be identified, and the study group was forced to make its recommendations on rationalization on a less firm statistical base than was desirable.

Services for children with cancer

The management of cancer in children requires special skills, and, in view of the low incidence of the disease, few hospitals see enough cases to develop optimal levels of care. Evidence from clinical trials suggests that children with cancer who are managed in specialist units have a better prognosis than those managed elsewhere (Lennox *et al.,* 1979; Kramer *et al.,* 1984). For a specialist unit to function optimally, however, the British Paediatric Association estimates that a minimum throughput of about 50 cases a year is required, and a maximum of around 100.

The overall annual incidence of cancer in the 1–14 age group in England and Wales is about 1000 cases. This suggests the need for between 10 and 20 units, or roughly one for each health service region. However, the regions vary considerably in population, and, at a national rate for childhood cancer of 2 per 100 000 total population (10 per 100 000 population in the 1–14 age group), some regional populations would give rise to less than the lower limit of 50 cases per year. These estimates, based on national data, have been confirmed through regional cancer registration data. The smallest region, East Anglia, with a total population of 1.9 million in 1981, registered 38 cases of cancer in children under age 15 in the same year. On the basis of these figures alone, this region would be unable to support a viable paediatric oncology service.

The needs of children with leukaemia and lymphoma are different from those with solid tumours, and different facilities are required for their management. Of the 38 children registered in East Anglia in 1981, 15 had leukaemia or lymphoma. This split further reduces the likely throughput for a regional unit and suggests the need for a small number of units specializing either in solid tumours of children or in the leukaemias and lymphomas, placed according to national incidence patterns. The policy for the development of paediatric oncology services therefore emphasized the need for interregional collaboration in planning and the sharing of resources.

Rare tumours

Some rare tumours, in children or in adults, have such particular requirements for optimal management that they are best managed in one or two national units. Wilms' tumour, retinoblastoma, germ-cell tumours of the testis and chorionepithelioma are a few examples. Cancer registration data indicate fairly precisely how many of these cases are to be expected each year, from what age groups and, in some cases, from which section of the population. This information is invaluable in planning a service to meet these patients' needs. Variations in incidence over the country may give an indication of where such specialist units could best be sited.

Bone-marrow transplantation

The advent of bone-marrow transplantation (BMT) as a technique for the management of bone-marrow dysfunction has considerable implications for health services because of its cost and the special skills required. As with any health care innovation, it is necessary to identify the need and to estimate the implications of introducing a service to meet the need as early as possible. The urgency in relation to BMT was enhanced by the wide publicity surrounding the development of the technique. The British Government set up an expert working party in 1982 to advise on the need for a service.

The availability of national incidence data for the leukaemias by pathological type and five-year age group made estimation of the numbers of possible candidates for BMT arising from this group of conditions relatively straightforward. On the basis of the current indications for BMT, it was possible to arrive at fairly precise figures for the numbers of cases occurring annually and to calculate from those the facilities and manpower required and the likely resource consequences. For the other conditions in which BMT might be indicated—aplastic anaemias and metabolic conditions—no incidence data were available, and estimates of numbers of cases were based on mortality data.

Having quantified the need, the pattern of service required to meet it has to be identified. This is determined by the availability of the necessary resources and expertise, rather than by any regional variation in potential case-load, since the incidence pattern of the leukaemias is fairly uniform.

Cancer screening services

The planning of large-scale cancer screening services, for cervical and breast cancer for example, requires the availability of registration data for assessing the extent and pattern of provision required and for evaluating the results (see Parkin & Day, this volume). Detailed planning at the subregional level requires registration information by district or local authority area.

CANCER REGISTRATION DATA IN EVALUATION OF SERVICES

On a national level, cancer registration data can give an overview of the changes in incidence and survival from which conclusions may be drawn about general trends and which enable the results of smaller-scale studies to be seen in context. Mortality data alone are of very limited usefulness in this context, and, as survival increases, information on mortality becomes even less useful. Changes in basic incidence are gradual, of complex causation and may relate to events occurring many years previously. Observation of such changes is therefore of limited usefulness in health service planning or evaluation. But there are instances in which monitoring of trends in incidence can be very useful, for instance when a population is at risk from exposure to carcinogens in the environment and in monitoring the results of environmental protection measures, of other preventive measures or of health education campaigns (see Teppo *et al.*, this volume).

Probably more useful, however, is the information on survival that can be obtained when cancer registry data can be linked to mortality data. While clinical trials are vital in defining in detail the effects of a particular treatment regimen on a particular group of patients, the effects of general advances in therapy on the outcome of a disease in the population as a whole are important in the evaluation of health services.

Analysis of survival data in the UK suggests that there has been a small improvement overall but that a marked improvement has occurred for some cancers, particularly Hodgkin's disease in men, the leukaemias (especially acute lymphoblastic leukaemia), the teratomas, cancer of the kidney in children, laryngeal cancer in men and oropharyngeal cancer. These improvements are probably due mainly to advances in treatment, although early diagnosis may play an important part in some cancers.

What can be done using national data is limited by a number of factors, not least of which is the variable quality of data from different parts of the country. The England and Wales national scheme does not call for details of stage at registration or treatment given; however, at the regional level, these data are often collected, and the variation in quality of data over time is much less and is more precisely known. Regional data may be used for evaluating progress in the management of cancer in that population, and a number of regional cancer services in England use these data to the full to assess the effectiveness of services and to help identify possible deficiencies.

REFERENCES

Cohen, C.S., Connolly, C.C., Matthews, G.G. & Skeet, R.G. (1982) Information needs for radiotherapy services. *Br. J. Radiol.*, **55,** B51–B54

Department of Health & Social Security (1978) *Health Circular (78)32,* London

Donnan, S. (1982) *Cancer registration—advance or retreat?.* In: Smith, A., ed., *Recent Advances in Community Medicine No. 2,* London, Churchill-Livingstone, pp. 157–168

Enterline, J.P., Parker, D.F. & White, J.E. (1982) Planning applied population-based cancer control programs: The uses of mortality and morbidity data. *Prog. clin. biol. Res.*, **83,** 207–217

Gandy, R.J. (1980) Cancer morbidity measurement. *Public Health (London)*, **94,** 40–43

Hakama, M. (1980) Projection of cancer incidence: Experiences and some results in Finland. *World Health Stat. Q.*, **33,** 228–240

Kay, H.E.M., Powles, R.L., Lawler, S.D. & Clink, H.M. (1980) Cost of bone marrow transplants in acute myeloid leukaemia. *Lancet*, ***i,*** 1067–1069

Kramer, S., Meadows, A.T., Pastore, G., Jarrett, P., Bruce, D. & Evans, A.E. (1984) Influence of place of treatment on diagnosis, treatment and survival in three pediatric solid tumors. *J. clin. Oncol.*, **2**, 917–923

Lennox, E.L., Stiller, C.A., Morris Jones, P.H. & Kinnier Wilson, L.M. (1979) Nephroblastoma: treatment during 1970–3 and the effect on survival of inclusion in the first MRC trial. *Br. med. J.*, ***ii***, 567–569

Office of Population Censuses and Surveys (1981) *Report of the Advisory Committee on Cancer Registration*, London, Her Majesty's Stationery Office

Pedersen, E. (1962) Some uses of the cancer registry in cancer control. *Br. J. prev. soc. Med.*, **16**, 105–110

Silman, A.J. & Evans, S.J.W. (1981) Regional differences in survival from cancer. *Community Med.*, **3**, 291–297

Standing Medical Advisory Committee, Standing Sub-Committee on Cancer (1984) *Report of the Working Group on Acute Services for Cancer*, London, Her Majesty's Stationery Office

Weddel, J.M. (1973) Registers and registries: A review. *Int. J. Epidemiol.*, **2**, 221–228

Wrighton, R.J. (1979) Cancer chemotherapy. *J. R. Soc. Med.*, **72**, 1–2

7. SURVIVAL RATE AS AN INDEX IN EVALUATING CANCER CONTROL

A. HANAI & I. FUJIMOTO

Center for Adult Diseases, Osaka Department of Field Research, Osaka Cancer Registry, Osaka, Japan

SUMMARY

A WHO/IARC Expert Committee on Cancer Statistics (1979) reported that it is of increasing importance to make comparisons of survival experience between countries. Unfortunately, there may be insufficient uniformity between different regions with regard to the criteria or classifications used and to registry policy and methods. The study of time trends in survival is complicated by changes in the degree of case reporting to registries, by the increasing number of cancer patients found by screening programmes, and by changes over time in definitions or coding rules of cancer site, stage, histology and even diagnosis. Such problems are inevitable in the medical field, which is evolving continually. Analysis of such data has, however, made important contributions in the field.

Survival rates present problems and limitations; however, the overall results from population-based cancer registries are more representative of the general pattern of survival from cancer than those from hospitals (WHO/IARC Expert Committee on Cancer Statistics, 1979). Interesting though a hospital patient series may be, it should be kept in mind that the real efficacy of a cancer control programme can be judged only from population figures.

INTRODUCTION

One of the purposes of population-based cancer registries is to evaluate the results of all activities for the prevention and medical care of cancer in a region. Assessment of the results of primary and secondary prevention of cancer is dealt with in other chapters. The possible uses and limitations of cancer registry survival data in the evaluation of the results of cancer care are considered here.

A CANCER REGISTRATION SYSTEM FOR THE EVALUATION OF CANCER CONTROL

Population-based cancer registries record data on all cases of cancer occurring in defined populations. The sources of information are diverse and usually include a number of different medical institutions. Collection of detailed information on site-specific diagnostic methods and treatment is difficult for many registries. Even when the registry adopts a widely recognized classification of the extent of a disease or a pathological classification approved by a national or international technical expert committee, not all reporting institutions may accept such definitions or use them consistently and reproducibly.

Because the data gathered are representative of entire populations in defined areas, population-based cancer registries should, in theory, be able to assess the results of comprehensive cancer care in those regions. The purpose and the subjects of such assessments then naturally differ from those of clinical trials, which, as the Commission on Cancer Control of the UICC (Knowelden *et al.*, 1970) pointed out, evaluate specific treatment methods for cancer patients in defined stages of the disease.

COLLECTION OF INFORMATION ON PATIENT PROGNOSIS

Active follow-up

Regional (population-based) cancer registries in the USA and in Canada collect follow-up information on patients from each reporting hospital cancer registry; these in turn conduct annual follow-up surveys of registered cancer cases through the patient's own doctor. This kind of survey is termed a 'medical follow-up'. The American College of Surgeons (1981) recommends that a follow-up system be implemented to ensure that the patient continues to see his/her physician for examination at regular intervals. With this kind of follow-up, therefore, the quality as well as duration of survival may be assessed.

Most population-based cancer registries in other countries, however, do not have a follow-up system for individuals such as is obtained in the USA or Canada. The subjects to be followed-up are all the cancer patients who are present in the regional population. Thus, a different, active follow-up method may be used to confirm the patient's survival or death indirectly by utilizing surveys or registries set up for other purposes. Many regional registries therefore use sources such as a population register (city directory), a comprehensive register for a national health service, a health insurance or social security register, voter lists, driving license register, etc. These techniques may also be used to trace the fate of cases lost during medical follow-up.

Passive follow-up

The principal procedure in the so-called passive follow-up method is to collect information on registered patients at the time of their death, using the death certificate file for the region. Collation of the two files—the death certificate file from vital statistics and the registry file of registered cases—is performed either in the cancer registry or in the local or national department of vital statistics. In the matching process, a combination of several indices (Hanai *et al.,* 1973) or national index numbers (Office of Population Censuses and Surveys, 1980; Hakulinen *et al.,* 1981) is used for patient identification.

In passive follow-up, any registered cancer patient whose death has not been notified to the registries by the department of vital statistics (in other words, all unmatched cases) are considered to be surviving. The result of passive follow-up may, therefore, be an overestimate of the true survival rate: the size of the error is due both to the accuracy of the matching process and to the emigration of registered cancer cases to other regions.

COMPUTING SURVIVAL RATE

Cumulative survival rate

The traditional method for computing cumulative survival rate is the actuarial or life-table method introduced by Cutler and Ederer (1958), although other methods were reported by Littel (1952) and Chiang (1961). Most population-based cancer registries have reported the end results for the registered patients as relative survival rates using the life-table method.

Subjects to be observed: When computing cumulative survival rate by the life-table method, two groups of patients are treated as censored cases—those patients lost to follow-up during the observation period and cases known to be alive but withdrawn from observation as they have not completed the time interval of interest. Besides these groups, there may be some patients for whom follow-up cannot be conducted because of lack of identification, such as full name or full address. Ries *et al.* (1983) excluded such persons rather than treating them as censored cases.

The risk of a cancer patient having multiple primaries has increased in recent years. Setting up a standard method for treating multiple cancers in computing survival is, therefore, necessary as well as devising rules for determining the incidence of cancer. As the prognosis of the second or subsequent primaries might be related to the results of treatment for the initial primary, Ries *et al.* (1983) excluded all primary cancers except the first in their analysis of survival in the SEER Program.

Cases registered without histological confirmation of diagnosis might include some non-cancerous patients, which would tend to make the survival rate higher. However, such cases often have more advanced diseases or receive unsatisfactory cancer care. In fact, it appears that they exhibit a lower survival than those confirmed histologically. Examples from Finland and Osaka are shown in Table 1. Lourie (1964)

Table 1. Five-year relative survival rate for lung cancer cases by method of diagnosis

Sex	Method of diagnosis	Five-year RSR[a] (%)	
		Finland[b] (1964–1974)	Osaka[c] (1970–1974)
Males	Histology	10.4	10.0
	Cytology	5.0	–
	All cases	–	5.8
Females	Histology	12.6	11.3
	Cytology	4.5	–
	All cases	–	6.6

[a] Relative survival rate
[b] From Hakulinen et al. (1981)
[c] From Fujimoto (1980)

suggested that data on histologically confirmed and unconfirmed cases should be tabulated separately, as well as in combination.

Cases registered by death certificate only: The proportion of cases registered by death certificate only (DCO) is an important factor to be considered when interpreting survival data from registries. The proportion of DCO cases is widely used as a measure of the completeness of case collection by a registry (Waterhouse et al., 1982). When this proportion is high, it may be assumed that some cases escape registration at the time of diagnosis, and, if they are completely cured, will never be recorded in incidence data. If this assumption is true, the registered cases would be a somewhat biased sample of all cancers, since there would be an excess of fatal cases, and survival rates computed for all registered (incident) cases would be lower in some degree than true survival.

DCO cases are often excluded from analyses (Griswold et al., 1955; Ries et al., 1983). Griswold et al. (1955) stated that 'they were interested in survival rates as a measure of success in case management'; they thus excluded the DCO cases (the proportion of which was 24.6% of all incident cases in Connecticut, 1935–1951). In these circumstances, however, computed survival might have a bias towards higher survival rates. Griswold et al. (1955) suggested that a sizeable portion of cancer cases registered on the basis of DCO was probably comprised of those who visited a physician during their terminal stage, and these physicians could do little or nothing to prolong the lives of such individuals. However, the exclusion of such cases means that survival rates no longer reflect the average prognosis of incident cases of cancer in the region, and the value of survival data as an index of the overall effectiveness of cancer care is considerably diminished.

The methods for handling these cases are limited because registries have only information that appears on death certificates. When computing incidence, the date of diagnosis for DCO cases is taken as the date of death (Griswold et al., 1955; Young et al., 1981). The duration of survival is thus considered to be zero. When computing

Table 2. Comparison of five-year survival rates (all cancers) by follow-up method, California and Osaka

	Follow-up method		
	Passive	Active	
		Lost cases assumed to be alive[a]	Lost cases treated as lost[b]
California (1969–1972)[c] (observed rate)	42.3	39.5	38.8
Osaka (1975–1977)[d] (relative rate)			
including DCO[e]	27.2	26.9	24.6
excluding DCO	33.8	33.5	31.7

[a] Included in effective number exposed to risk of dying during the year
[b] Computed by the normal actuarial method
[c] From Austin (1983)
[d] From Hanai (unpublished data)
[e] Cases registered with death certificate only

cumulative survival by the life-table method, such individuals have to be included among patients surviving less than one year. As a result, the one-year survival rate is unavoidably and artificially reduced.

Duration of disease is included as an obligatory item on the death certificate form in some countries; in such circumstances, it can be used in calculations of survival as the interval between diagnosis and death. Hanai et al. (1978) used this method to assess the extent of artefact introduced by assigning DCO cases a survival of zero. As the duration of disease for DCO cases was almost always found to be less than five years, an artificial decrease remained in three-year cumulative survivals, but not in five-year survivals.

It might be reasonable to report two survival rates—one for incident cases, including the DCO cases, and the other for reported cases, excluding them. Lourie (1964) also suggested using all available data. In any case, the proportion of DCO cases should be stated in the survival report.

Cases lost during active follow-up: Cases lost during active follow-up are treated as censored cases in the life-table method. One half of the lost cases are subtracted from the denominator when computing mortality during the year. However, it is probable that a greater proportion than this are still alive. Austin (1983) concluded from routine follow-up experience at the California Tumor Registry that 'the traditionally computed survival rates are too pessimistic and underestimate the true rates'. Assuming all lost cases to be alive, the survival rate rises a little—as shown in Table 2. In the follow-up, the percentage of lost cases should be kept as small as possible.

Adjusted survival rate

All deaths occurring among the subjects, including deaths from causes other than cancer, are used in the calculation of cumulative survival rates. If reliable information is available on the cause of death of all registered cancer patients, 'adjusted' survival rates can be obtained by withdrawing cases for which cause of death was other than the registered cancer. The American Joint Committee (1977) pointed out that the method is useful for excluding the effect of factors such as sex, age, race and socio-economic status on survival rates. However, information on cause of death is often unreliable or insufficient.

Relative survival rate

It is frequently difficult to classify the cause of death of cancer patients into a cancer death or non-cancer death. Sometimes such classification is subject to unconscious bias. In order to exclude this problem, relative survival rate uses only the fact of death, irrespective of cause. The expected survival of the patient group is computed on the basis of mortality of the whole population of the registration area and compared with the observed cumulative survival rate of the patients. The ratio of the observed survival rate to the expected rate is called the 'relative survival rate'. The relative survival rate is thus that corrected for the probability of death from other causes given by age and sex. It indicates the excess risk of death due to cancer.

For most cancer sites, the observed cumulative survival curve of patients becomes parallel with the expected survival curve computed from mortality of the general population five years after initial treatment. Conventionally, therefore, cancer cases are considered to be cured if no recurrence is noted within five years after treatment. [An exception is cancer of the breast, from which excess mortality is seen in cases for many years following treatment (Brinkley & Haybittle, 1984).] Thus, five-year relative survival rates are commonly used in assessing cancer treatment.

USES OF SURVIVAL RATES FROM POPULATION-BASED CANCER REGISTRIES

Problems in the interpretation of survival

From 1959, the US National Cancer Institute was concerned with a cooperative international effort to evaluate cancer therapy (Cutler, 1964). Before collecting survival data from different countries, exchange visits were conducted between the USA and Europe to investigate differences in the definitions of registered items and methods in individual registries. After the visits, Lourie (1964) reported that comparisons of survivals could be affected by completeness of reporting, definition of stage and treatment, follow-up methods etc. Logan (1978) and Doll and Peto (1981) referred to the differences in standards of diagnosis as well as to the factors mentioned above.

In theory, survival rates from a population-based cancer registry should be a useful index for evaluating comprehensive cancer care in a region and not simply a reflection of the results of treatment (Morrison *et al.*, 1976). Survival rates should be understood as a composite index which expresses the results of treatment for the cancer, including significant prognostic factors such as stage at diagnosis, histological type and other characteristics of the disease. Besides these, genetic, somatic and socio-economic conditions of the patient groups may account for possible differences in survival rates, even after they are adjusted by stage (Berg *et al.*, 1977).

When the definition of such prognostic factors within a single category of cancer (defined by site) differs between hospitals or registries, there are difficulties in making valid inferences from comparisons of survival rates.

In the following sections, recently published data from several population-based cancer registries are compared, although the problems in interpretation discussed above should be borne in mind.

Intercountry comparison of survival rates

At an international symposium in Norway in 1963, and ad-hoc group first presented population-based survival data from six countries: Denmark, England and Wales, Finland, France, Norway and the USA (Haenszel, 1964).

In 1978, Logan (1978) reviewed cancer survival statistics published by six population-based registries: England and Wales, France, Norway, Poland, Scotland and the USA. Two sets of data communicated to the World Health Organization from the German Democratic Republic and Osaka (Japan) were reviewed as well.

Table 3 shows five-year relative survival rates of seven populations by site and sex, collected from recent publications of six registries or unions of registries in various countries. Lymphoma and leukaemia were not listed because the classification was different for each registry. For the Osaka Cancer Registry, two survival rates are given, one for incident cases (including DCO cases) and the other for reported cases (excluding DCO cases). The proportion of DCO cases in Osaka during 1975–1977 was 24.7% for all sites in both sexes.

Differences of completeness of reporting would affect the comparability of survival rates (Lourie, 1964; Logan, 1978; Doll & Peto, 1981). The proportion of DCO cases is rarely referred to in publications of survival data. As shown in *Cancer Incidence in Five Continents Vol. IV* (Waterhouse *et al.*, 1982), the proportion is negligibly small in most registries listed in Table 3 (other than Osaka), and no information is available for a few registries.

The survival rates were low for cancers of the oesophagus, liver, pancreas and lung (0–11%), stomach (7–25%) and ovary (18–37%). However, they were higher for cancers of the larynx, prostate and bladder (32–75%), breast and uterine cervix (54–87%). The proportionate distribution of cancer sites was quite different between these populations; thus, survival for all sites was excluded from the table.

Logan (1978) reported large differences in survival for cancer at certain sites from the international data. He suspected that the criteria for a malignant tumour differed between registries. The lowest survival figures in the series in Table 3 were improved in comparison with Logan's series of patients from 1955–1969 for every site of cancer except cancers of the oesophagus, liver, pancreas and lung, which still show the poorest survival rates.

The US white population showed the highest survival for most cancer sites. The highest survival for stomach cancer was in Osaka; for cancers of the

Table 3. Comparison of five-year survival rates for cancer patients by sex and site in seven population in six countries

Site	Sex	Five-year relative survival rate (%)							
		Japan Osaka[a]		USA SEER[b]		England & Wales[c]	Norway[d]	Finland[e]	Iceland[f]
		DCO cases[g] incl. (1975–1977)	excl.	White (1973–1979)	Black	(1971–1973)	(1971–1973)	(1972–1975)	(1970–1974)
Oesophagus	M	5.7	7.8	–	–	6.3	3	3.8	8
	F	7.7	11.2	–	–	7.9	7	5.2	0
Stomach	M	24.2	30.9	12	13	7.4	14	9.5	12
	F	22.4	28.6	14	16	7.3	14	8.3	11
Colon	M	27.3	34.7	47	41	29.6	40	32.3	32
	F	25.6	34.6	49	46	29.4	41	31.2	29
Rectum	M	24.6	31.7	44	28	30.8	36	28.3	25
	F	22.8	29.7	47	41	32.9	40	32.1	38
Liver	M	1.0	1.4	–	–	2.3	5	1.5	0
	F	2.1	2.6	–	–	3.0	9	2.3	0
Pancreas	M	1.5	5.2	3	3	3.8	2	1.6	0
	F	3.3	4.6	2	6	3.1	3	2.3	2
Larynx	M	56.7	63.3	–	–	64.4	75	57.3	36
	F	50.9	57.6	–	–	56.9	48	62.0	33
Lung	M	6.2	8.1	10	8	7.8	7	7.4	6
	F	6.9	9.1	14	11	7.0	9	9.8	11
Breast	F	64.8	69.6	72	60	56.8	68	57.6	59
Uterine Cervix	F	56.1	62.5	66	61	54.4	73	57.9	69
Uterine Corpus	F	57.6	60.7	87	54	–	79	59.5	63
Ovary	F	18.2	24.2	34	35	25.3	37	32.2	–
Prostate	M	32.8	38.8	64	54	35.9	52	40.1	36
Bladder	M	46.0	54.5	72	49	53.8	40	41.2	53
	F	34.8	45.8	69	34	47.4	32	39.8	55

[a] From Hanai et al. (unpublished data)
[b] From Ries et al. (1983)
[c] From Office of Population Censuses and Surveys (1980)
[d] From Magnus (1980)
[e] From Hakulinen et al. (1981)
[f] From Bjarnason & Tulinius (1983)
[g] Proportion of the cases with death certificate only (DCO) is 26.5% for males and 22.7% for females

Table 4. Proportion of localized cases and their five-year relative survival rates[a]

Site	Sex	Japan Osaka (1975–1977)			USA SEER (1964–1973)				Norway (1968–1975)		Finland (1953–1974)	
		A		B	White		Black		A	B	A	B
		DCO cases			A	B	A	B				
		incl.	excl.									
Stomach	M	12.9	17.6	69.1	13	48	14	47	25	38	22.5	26.6
	F	11.7	16.4	67.5	16	49	15	48	23	34	23.5	22.5
Colon	M	13.0	18.7	76.0	35	77	30	66	37	68	34.1	59.3
	F	11.6	19.8	65.4	33	80	29	61	36	69	35.3	59.5
Lung	M	12.0	16.9	28.3	13	35	15	19	33	19	34.9	13.9
	F	10.2	14.6	21.3	15	60	13	44	30	26	27.4	24.3
Breast	F	32.6	41.1	91.6	47	88	32	79	45	86	45.4	77.1
Uterine cervix	F	33.3	35.9	86.7	44	82	38	78	47	89	69.8	66.9

[a] Sources as in footnote to Table 3; except for USA, from Myers & Hankey (1980)
A, % of cases 'localized; B, five-year relative survival rate (%)

liver[1], larynx (male)[1], uterine cervix and ovary in Norway; and for laryngeal cancer[1] (female) in Finland. The reported cases (excluding DCO cases) in Osaka showed similar survival to those in England and Wales and in Finland.

It is possible that a difference in survival rates between two areas is due to a different stage distribution, or to a difference of survival within stages.

The proportions and survival of localized cases were available for five populations from four registries. Published data were available for 1964–1973 for the SEER registries in the USA. Table 4 shows the proportions and the survival rates of localized cases for the five most frequently occurring sites. When the proportion of localized cases and their survival rates from this table are multiplied, the result correlates highly with survival rates for all cases at the same site (correlation coefficients between 0.6 and 0.9).

Table 5 shows the ratio of survival of localized cases to that of all cases. In general, the ratio is small for the sites showing relatively long survival and large in the sites with poor survival.

Intracountry comparisons of survival rate

Intercountry comparisons of survival rates are difficult because of variations in definitions and in methods between registries. These problems are less evident in

---

[1] For these sites, survival data for the USA were not available and could not be compared with data for the other populations.

Table 5. Ratio of five-year relative survival rate (RSR) for localized cases to that for all cases in five populations

Site	Sex	Ratio of five-year RSR (localized cases/all cases)				
		Japan Osaka[a] (1975–1977)	USA SEER[b]		Norway[c] (1968–1975)	Finland[d] (1953–1974)
			White (1973–1979)	Black		
Stomach	M	2.9	4.4	3.6	2.7	2.8
	F	3.0	3.5	4.0	2.8	2.7
Colon	M	2.8	1.7	1.9	1.8	1.8
	F	2.6	1.6	1.7	1.7	1.9
Lung	M	4.6	4.8	3.8	2.4	1.9
	F	3.1	4.6	4.9	2.4	2.5
Breast	F	1.4	1.3	1.6	1.3	1.3
Uterine cervix	F	1.5	1.4	1.5	1.2	1.2

[a] From Hanai et al. (unpublished data)
[b] From Myers & Hankey (1980)
[c] From Magnus (1980)
[d] From Hakulinen et al. (1981)

comparisons of survival between different populations within a registry area or between registries within a region where the same methods of registration are in use.

Table 6 presents survival data for two series: from six local areas in Osaka (1975–1977) and from 15 regions in England and Wales (1971–1973). In discussing the latter data, Silman and Evans (1981) noted that survival from the more lethal cancers had the largest relative differences and less lethal cancers the largest absolute differences. The same was observed in the series from Osaka.

The ratio of the highest survival to the lowest was calculated and found to be smaller (1.1–1.2) for cancer sites with better survival, like breast or uterine cervix, and greater in the sites showing unfavourable survival rates, like lung (2.3–2.6). Comparing the two registries, it was interesting that for stomach and colon cancers, the ratios were smaller in the registry with a higher survival rate as well. Local areas showing high survival rates for one cancer site tended to show the same for many other cancer sites, and the areas with low survival rates for one cancer site tended to show poor survival rates for other cancer sites as well.

A proportion of cancer patients will seek better medical care in specialized institutions outside their residential area. Therefore, assessment of patient movement is necessary when analysing survival differences by area. Such a study was conducted in Osaka (Fujimoto, 1980; Hanai, 1981). The area covered by the registry has a population of 8.5 million. It consists of Osaka city, with a population of 3 million, and five surrounding areas. Half of the medical doctors, hospitals and hospital beds

Table 6. Difference by local area of five-year survival rates of cancer patients for selected sites in Osaka[a] and in England and Wales[b]

Site	Sex	Five-year relative survival rate (%)				Ratio of highest to lowest	
		Japan Osaka six areas (1975–1977)		England & Wales 15 areas (1971–1973)		Japan Osaka six areas (1975–1977)	England & Wales 15 areas (1971–1973)
		Lowest	Highest	Lowest	Highest		
Stomach	M	18.4 (Sensyu)	27.5 (Sakai)	4.4 (Manchester)	11.6 (NWM)	1.5	2.6
	F	16.3 (Sensyu)	25.7 (HMD)	3.3 (Liverpool)	11.8 (NWM)	1.6	3.6
Colon	M	18.0 (Sensyu)	36.7 (HMD)	22.6 (Manchester)	34.8 (NWM)	2.0	1.5
	F	11.9 (Sensyu)	32.7 (HMD)	23.4 (Newcastle)	34.8 (NWM)	2.7	1.5
Lung	M	3.1 (Sensyu)	8.1 (Sakai)	5.2 (Leeds)	12.0 (NWM)	2.6	2.3
	F	4.0 (HMD)	9.0 (S. Kawachi)	4.5 (Leeds)	10.4 (NWM)	2.3	2.3
Breast	F	60.3 (Sensyu)	71.6 (Sakai)	53.8 (Liverpool)	61.3 (SWM)	1.2	1.1
Uterine cervix	F	49.7 (Sakai)	64.1 (S. Kawachi)	49.7 (Newcastle)	61.3 (Wales)	1.2	1.2

Local districts in parentheses: NWM, North-west metropolitan; SWM, south-west metropolitan; HMD, Honō Mishima District; S. Kawachi, southern Kawachi
[a] From Hanai *et al.* (unpublished data)
[b] From Office of Population Censuses and Surveys (1980)

in Osaka Prefecture are concentrated in Osaka city. Among cancer patients in the five surrounding areas, younger persons and those in the earlier clinical stages were more likely to visit medical institutions located in Osaka city. It was observed for each surrounding area that the cancer cases diagnosed and treated in Osaka city received more thorough medical examinations and were more often treated surgically than the patients who remained inside their residential area. Consequently, survival was higher in the group visiting institutions outside their residential areas. As the proportion of such persons is different by local area, the overall effectiveness of medical care in a given area was studied by evaluating both the medical care received by resident patients in the area and the care received by residents who left the area to receive treatment elsewhere.

Comparison of survivals by age

Survival rate by ten-year age group was available from three registries: Osaka, USA and England and Wales. The survival rates for four 10-year age groups are illustrated in Figure 1 for five major sites. In the figure, the data for stomach, colon and lung cancers are given only for males and those for breast and cervix for females. The US data are for whites only. In the USA, a clear age differences in relative survival is seen only for cancers of the lung and uterine cervix. In Osaka and England and Wales, however, survival rates decreased with age for each site considered. For carcinoma of the cervix, the Osaka data (1975–1977) show equivalent variations in stage distribution and treatment rate by age. Stomach cancer in Osaka shows little variation in the percentage of localized cases by age (range, 21%–16%), which could explain survival differentials. However, the proportion of cases of unknown stage increased with age (25% to 37%). In addition, there was a considerable difference in the proportion of cases that were surgically operated: 81% under age 54, but 31% over age 75. Hanai *et al.* (1982a) reported that the intestinal type of gastric carcinoma was more prevalent in older age groups and the diffuse type more common at younger ages (shown in Table 11). The survival rate in Osaka was found to be higher for the intestinal type than for the diffuse type of cancer. If there is no other difference between the older and younger patient groups, survival would thus be higher in the former group. Since the converse was observed, it was considered that older people found it more difficult to receive good medical care in the early stages of their disease.

Comparison of survival rates by sex

The sex ratio of survival rates is listed by site in Table 7. For cancers of the oesophagus, liver, rectum and lung, sex ratios were commonly lower than 1.0, indicating better survival in females than in males except for cancer of the rectum in Japan and of the lung in England and Wales. For laryngeal cancer the sex ratio was over 1.0, except for Finland, which indicates better survival experience in males than in females.

It has been reported (Wynder *et al.*, 1956; Vincent *et al.*, 1965) that the histological distribution of lung cancer is different between the two sexes, adenocarcinoma being much more prevalent among females. The same was found in Osaka (Hanai *et al.*, 1982b) and Finland (Hakulinen *et al.*, 1981), as shown in Table 8. This table also shows higher survival rates in Finland for patients with adenocarcinoma than for those with other types of lung cancer. The more favourable survival rate in female lung cancer patients might result from the higher frequency of adenocarcinoma. However, since 1975 the surgical operation rate for squamous-cell carcinoma of the lung has markedly increased. This is reflected in the improvement in survival for this type of carcinoma. The present sex ratio may diminish in the near future.

An analysis of registry data (in 1970–1974) for bladder cancer in Osaka (Hanai & Fujimoto, 1983) indicated that the unfavourable survival of female patients was due mainly to delay in visiting medical institutions. For oesophageal cancer, they reported that the sex difference in survival remained even for surgically operated patients with the same extent of disease. They reported that the etiological factors of oesophageal cancer, especially for males, might adversely affect prognosis.

Fig. 1. Comparison of five-year survival rates of patients with cancers at selected sites, by age group in three populations: Japan (Osaka), 1975–1977 (Hanai, unpublished data); USA (SEER Program), 1973–1979 (Ries et al., 1983); England and Wales, 1971–1973 (Office of Population Censuses and Surveys, 1980)

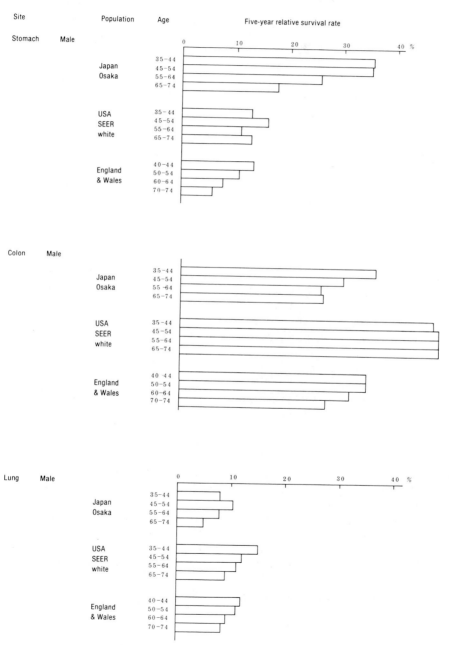

Fig. 1 (contd)

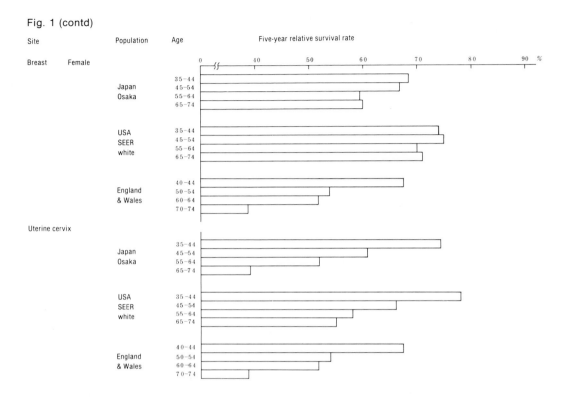

Comparison of survival rates by racial group

Table 9 shows the survival rates of white and black cancer patients in the USA from the SEER series in 1964–1973 (Myers & Hankey, 1980). The survival rates for all stages and for the localized cases alone are shown, together with the ratio of survival of whites *versus* blacks. The ratio was higher than 1.0 for every site except stomach cancer in males. Some of the difference may be ascribed to delay in diagnosis, since the percentage of localized cases is lower in blacks than whites (Table 4). However, there is still a difference in survival, even though reduced, when only localized cases are compared.

The racial difference has been ascribed to differences in socio-economic status (Dayal *et al.*, 1982), and these differences have also been suggested as the possible reason for some of the international variations (Berg *et al.*, 1977). However, a recent study (Chirikos *et al.*, 1984) did not find that economic status was an important variable when other factors were adequately controlled for in the analysis.

Further observations are necessary to define whether the variation lies in hereditary differences between the two races, or whether it comes from small differences in the extent of the disease within the range of the same clinical stage.

Table 7. Sex difference of five-year survival rates for cancer patients by site in seven populations[a]

Site	Sex ratio of survival (males:females)						
	Japan Osaka (1975–1977)	USA SEER		England & Wales (1971–1973)	Norway (1972–1975)	Finland (1953–1974)	Iceland (1970–1974)
		White (1973–1979)	Black				
All sites	0.6	0.7	0.6	0.7	–	0.6	0.6
Oesophagus	0.7	–	–	0.8	0.4	0.7	–
Stomach	1.1	0.9	0.8	1.0	1.0	1.1	1.0
Colon	1.1	1.0	0.9	1.0	1.0	1.0	1.1
Rectum	1.8	0.9	0.7	0.9	0.9	0.9	0.7
Liver	0.5	–	–	0.8	0.6	0.7	–
Pancreas	0.5	1.5	0.5	1.2	0.7	0.7	–
Larynx	1.1	–	–	1.1	1.6	0.9	1.1
Lung	0.9	0.7	0.7	1.1	0.8	0.8	0.5
Bladder	1.3	1.0	1.4	1.1	1.3	1.0	1.0

[a] Sources as given in footnote to Table 3

Table 8. Proportions and survival rates of patients with different histological types of lung cancer

Sex	Histology	Proportions of cancers (%)		Five-year relative survival rate (Finland, 1967–1974)	
		Osaka[a] 1967–1972	Finland[b] 1967–1974	All stages	Localized
Male	Adenocarcinoma	34.6	9.5	16.9	42.5
	Epidermoid carcinoma	63.5	43.7	14.7	26.3
	Small-cell carcinoma	5.9	17.4	3.3	9.5
	Other carcinoma	12.1	29.4	6.4	16.1
Female	Adenocarcinoma	56.4	30.0	16.2	60.7
	Epidermoid carcinoma	29.5	20.2	10.9	16.9
	Small-cell carcinoma	5.1	16.9	5.1	18.2
	Other carcinoma	9.0	32.9	8.3	36.0

[a] From Hanai et al. (1982b)
[b] From Hakulinen et al. (1981)

Time trends in survival

The interpretation of time trends in survival within an area is complicated by the same problems of comparability that are encountered in studies between different areas. There are several particular problems with regard to time trends. Firstly, the reporting rate in many registries may not be satisfactory in the early period of operation. Logan (1978) and Doll and Peto (1981) suggested that improvements in survival rates could be caused partly by the better ascertainment and recording of incident cases. Secondly, when there has been a trend towards earlier diagnosis

Table 9. Comparison of five-year survival rates for cancer patients in the USA (1964–1973)[a]

Site	Five-year relative survival rate (%)											
	All stages						Localized					
	Males			Females			Males			Females		
	White	Black	(W/B)	White	Black	(W/B)	White	Black	(W/B)	White	Black	(W/B)
Stomach	11	13	(0.8)	14	12	(1.2)	48	47	(1.0)	49	48	(1.0)
Colon	46	34[b]	(1.4)	50	35[b]	(1.4)	77	66	(1.2)	80	61[b]	(1.3)
Lung	8	5[b]	(1.6)	13	9[b]	(1.4)	35	19[b]	(1.8)	60	47	(1.3)
Breast	–	–		68	49[b]	(1.4)	–	–		88	79[b]	(1.1)
Uterine cervix	.	.		59	51[b]	(1.2)	.	.		82	78	(1.1)

[a] From Myers & Hankey (1980)
[b] Statistically significant black-white differences at the 0.05 level (two-tail test)

(possibly as a result of the introduction of screening programmes), five-year survival may be improved, although the gain is due entirely to increased lead time, and mortality rates will remain constant (see Parkin & Day, this volume).

Time trends of survival rates in the early 1960s and in the early 1970s are shown in Table 10 for four populations[1]. Survival rates have been rising for each cancer site in all four populations during the decade, with few exceptions. The ratio of the more recent survival rates to the previous rate tends to be greater when previous survival had been comparatively low.

In Osaka, the survival rate for stomach cancer improved significantly during the period. A decrease in the frequency of the intestinal type of carcinoma (Table 11) had been expected to reduce the survival rate of all stomach cancer patients, but there was an increase in the proportion of patients receiving examination and treatment, as shown in figure 2A, and this was considered to have contributed greatly to the rise in survival above the expected rate (Hanai & Fujimoto, 1982).

Conversely, the survival rate for lung cancer patients in Osaka showed no improvement during the observation period (1963–1980). The proportions of patients who received endoscopy, cytology and histology (Fig. 2B) increased, especially in later years. However, the proportion of patients who received surgical operations during the same period showed no increase.

As seen in Table 10, the improvement in survival rates from cancers at the major sites is rather limited. However, dramatic improvements in survival have been reported for certain childhood cancers—lymphatic leukaemia, Hodgkin's disease, Wilms' tumour, etc. Table 12 shows data from the UK for the period 1962–1975. Similar results of a decrease in mortality from childhood cancer have been reported in the USA (Miller & McKay, 1984).

---

[1] For Osaka, only data from the late 1960s are available, since the registry started in 1963.

Table 10. Time trends in five-year relative survival rates for selected sites in four populations

Site	Sex	Japan Osaka[a]			USA, SEER[b] (White)			England & Wales[c]			Iceland[d]		
		1965–1969	1970–1974	(Ratio)[e]	1960–1963	1970–1973	(Ratio)	1962–1963	1971–1973	(Ratio)	1960–1964	1970–1974	(Ratio)
Stomach	M	18.3	20.9[f]	(1.1)	10	12	(1.2)	5.8	7.4	(1.3)	8	12	(1.5)
	F	15.9	16.6[f]	(1.0)	13	14	(1.2)	5.4	7.3	(1.4)	4	11	(2.8)
Colon	M	19.8	24.8[f]	(1.3)	42	47[f]	(1.1)	26.4	29.6	(1.1)	37	32	(0.9)
	F	17.0	19.4	(1.1)	44	50[f]	(1.1)	26.2	29.4	(1.1)	25	29	(1.2)
Lung	M	5.8	5.8	(1.0)	7	9[f]	(1.3)	5.9	7.8	(1.3)	11	6	(0.5)
	F	6.5	6.6	(1.0)	11	14[f]	(1.3)	4.8	7.0	(1.5)	0	11	(–)
Breast	F	57.9	60.3[f]	(1.0)	63	68[f]	(1.1)	46.6	56.8	(1.2)	46	59	(1.3)
Uterine cervix	F	48.7	52.7[f]	(1.1)	58	64[f]	(1.1)	46.7	54.4	(1.2)	33	69	(2.1)

[a] From Fujimoto (1980)
[b] From Myers & Hankey (1980)
[c] From Office of Population Censuses and Surveys (1975, 1980)
[d] From Bjarnason & Tulinius (1983)
[e] Ratio of the survival rate in the most recent period to that in the previous period
[f] Increase in survival was tested in the source text as statistically significant at the 0.05 level

Table 11. Five-year relative survival rates (RSR) and trend in the proportion of patients with carcinoma of the stomach by histological group in Osaka[a]

Sex	Histological group	Five-year RSR (%)	Proportion (%)[b]	
		1966–1977	1966–1971	1972–1977
Male	Intestinal type	52.7	57.8	54.3
	Diffuse type	39.9	41.2	44.5
	Other type	20.6	1.0	1.2
Female	Intestinal type	49.4	43.8	38.1
	Diffuse type	33.2	54.9	60.7
	Other type	23.5	1.3	1.2

[a] From Hanai & Fujimoto (1982)
[b] Proportion of patients for whom a histological classification was available

Fig. 2. Time trends in the percentage of cases admitted to hospital, undergoing diagnostic examination and treatment in Osaka, Japan (Fujimoto, unpublished data)

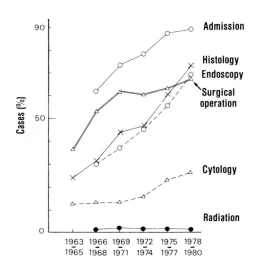
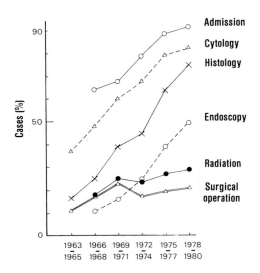

Table 12. Time trends of five-year relative survival rates for childhood cancers (0–14 years) in the UK[a]

Diagnosis	Five-year relative survival rate (%)	
	1962–1964	1971–1974
Lymphoid leukaemia	14[b]	39
Myeloid etc[c] leukaemia	4[b]	5
Unspecified leukaemia	10[b]	10
Hodgkin's disease	39	79
Non-Hodgkin's lymphomas	20	25
Neuroblastoma	18	15
Wilms' tumour	26	57
Retinoblastoma	85	88
Medulloblastoma	20	24
Ependymoma	16	36
Astrocytoma	54	56
Other and unspecified CNS[d]	28	31
Malignant bone tumours	22	29

[a] From Draper et al. (1982)
[b] Data for 1968–1970
[c] 'Myeloid etc' includes monocytic, myeolomonocytic and erythroid
[d] Including a few non-central nervous system intracranial tumours

Survival rates of selected patients

Population-based cancer registries register all cancer patients in an area in all clinical stages and undergoing all forms of treatment, both satisfactory and unsatisfactory.

Markedly high survival rates are sometimes observed in some groups, such as those detected by screening programmes and patients who had been diagnosed and received cancer treatment at cancer centres. The Japanese Research Society of Gastric Cancer (1983) collected 16 013 stomach cancer cases from 56 institutions in Japan from 1969–1973. Of these, 11 845 were resected cases and had 56.5% five-year relative survival rates[1]. Such favourable figures should be a target at which cancer medical care should aim for all cancer cases in the region.

REFERENCES

American College of Surgeons (1981) *Cancer Program Manual,* Chicago

American Joint Committee on Cancer Staging and End-Results Reporting (1977) *Manual for Staging of Cancer 1977,* Chicago

Austin, D.F. (1983) Cancer registries: A tool in epidemiology. In: Lilienfeld, A.M., ed., *Reviews of Cancer Epidemiology,* Vol. 2, New York, Elsevier, pp. 118–140

Berg, J.W., Ross, R. & Latourette, H.B. (1977) Economic status and survival of cancer patients. *Cancer, 39,* 467–477

Bjarnason, O. & Tulinius, H. (1983) Cancer registration in Iceland 1955–1974. *Acta pathol. microbiol. immunol. scand., 91,* Section A, Suppl. 281

Brinkley, D. & Haybittle, J.L. (1984) Long term survival of women with breast cancer. *Lancet, i,* 1118

Chiang, C.L. (1961) A stochastic study of the life table and its applications: the follow up study with the consideration of competing risks. *Biometrics, 17,* 57–78

Chirikos, T.N., Reiches, N.A. & Moeschberger, M.L. (1984) Economic differentials in cancer survival: A multivariate analysis. *J. chronic Dis., 37,* 183–193

Cutler, S.J. (1964) Introduction. *Natl Cancer Inst. Monogr., 15,* 5–20

Cutler, S.J. & Ederer, F. (1958) Maximum utilization of the life table method in analyzing survival. *J. chronic Dis., 8,* 699–712

Dayal, H.H., Power, R.N. & Chiu, C. (1982) Race and socio-economic status in survival from breast cancer. *J. chronic Dis., 35,* 675–683

Doll, R. & Peto, R. (1981) *The Causes of Cancer,* Oxford, Oxford University Press

Draper, G.J., Birch, J.M., Bithell, J.F., Kinnier Wilson, L.M., Leck, I., Marsden, H.B., Morris Jones, P.H., Stiller, C.A. & Swindell, R. (1982) *Childhood Cancer in Britain, Incidence, Survival and Mortality (Studies on Medical and Population Subjects No. 37),* London, Her Majesty's Stationery Office

Fujimoto, I. (1980) Meaning and utilization of population-based cancer registry [in Japanese]. *J. Osaka med. Assoc., 173,* 55–71

---

[1] Total resected cases   11845: 56.5% five-year relative survival

   Curatively resected cases   8531: 71.7% five-year relative survival
   Non-curatively resected cases   3245: 13.0% five-year relative survival
   Unknown   69: 47.4% five-year relative survival

Griswold, M.H., Wilder, C.S., Cutler, S.J. & Pollack, E.S. (1955) *Cancer in Connecticut, 1935–1951,* Hartford, Connecticut State Dept of Health

Haenszel, W.M. (1964) Contributions of end results data to cancer epidemiology. *Natl Cancer Inst. Monogr., 15,* 21–33

Hakulinen, T., Pukkala, E., Hakama, M., Lehtonen, M., Saxén, E. & Teppo, L. (1981) *Survival of Cancer Patients in Finland in 1953–1974 (Annals of Clinical Research, 13, Suppl. 31),* Helsinki, Finnish Cancer Registry

Hanai, A. (1981) *Uses of cancer survivals in population-based cancer registry* [in Japanese]. In: Fujimoto, I. & Oshima, A., eds, *Cancer Registry and Clinical Epidemiology,* Tokyo, Shinohara-Shuppan, pp. 118–133

Hanai, A. & Fujimoto, I. (1982) Cancer incidence in Japan in 1975 and changes of epidemiological features for cancer in Osaka. *Natl Cancer Inst. Monogr., 62,* 3–7

Hanai, A. & Fujimoto, I. (1983) *Sex differentials in survival rates of cancer patients in Osaka.* In: Lopez, A.D. & Ruzicka, L.T., eds, *Sex Differentials in Mortality, Trends, Determinants and Consequences,* Canberra, Australian National University, pp. 371–386

Hanai, A., Sakagami, F. & Fujimoto, I. (1973) Computerized cancer registration collation system. *Ann. Rep. Center Adult Dis., 13,* 1–10

Hanai, A., Oshima, A. & Fujimoto, I. (1978) Methods for computing survival rates in a population-based cancer registry and the reliability [in Japanese]. *Jpn. J. public Health, 25,* 485

Hanai, A., Fujimoto, I. & Taniguchi, H. (1982a) *Trends of stomach cancer incidence and histological types in Osaka.* In: Magnus, K., ed., *Trends of Cancer Incidence,* Washington DC, Hemisphere Publishing Corp., pp. 143–154

Hanai, A., Fujimoto, I., Hiyama, T. & Tateishi, R. (1982b) *Trends of lung cancer incidence and their histological distribution in Osaka.* In: Proceedings 13th International Cancer Congress Seattle, 1982. pp. 386

Japanese Research Society of Gastric Cancer (1983) *Report of the Gastric Cancer Registration Programme, No. 14,* Tokyo, National Cancer Institute, p. 124

Knowelden, J., Mork, T. & Philipps, A.J. (1970) *The Registry in Cancer Control (UICC Technical Report Series Vol. 5),* Geneva, International Union Against Cancer

Littel, A.S. (1952) Estimation of the T-year survival rate from follow-up studies over a limited period of time. *Human Biol., 24,* 87–116

Logan, W.P.D. (1978) Cancer survival statistics, international data. *World Health Stat. Q., 31,* 62–73

Lourie, W.I., Jr (1964) Some observations concerning comparability in these data. *Natl Cancer Inst. Monogr., 15,* 369–380

Magnus, K. (1980) *Survival of Cancer Patients, Cases Diagnosed in Norway, 1968–1975,* Oslo, The Norwegian Cancer Registry

Miller, R.W. & McKay, F.W. (1984) Decline in US childhood cancer mortality— 1950 through 1980. *J. Am. med. Assoc., 251,* 1567–1570

Morrison, A.S., Lowe, C.R., MacMahon, B., Ravnihar, B. & Yuasa, S. (1976) Some international differences in treatment and survival in breast cancer. *Int. J. Cancer, 18,* 269–273

Myers, M.H. & Hankey, B.F. (1980) *Cancer Patient Survival Experience (NIH*

Publication No. 80–2148), Bethesda, MD, US Department of Health and Human Services

Office of Population Censuses and Surveys (1975) *The Registrar General's Statistical Review of England and Wales 1968–1970 Supplement on Cancer*, London, Her Majesty's Stationery Office

Office of Population Censuses and Surveys (1980) *Cancer Statistics Survival, 1971–73 Registration, England and Wales (Series MB1, No. 3)*, London, Her Majesty's Stationery Office

Ries, L.G., Pollack, E.S. & Young, J.L., Jr (1983) Cancer patient survival: Surveillance, Epidemiology and End Results Program, 1973–79. *J. natl Cancer Inst., 70,* 693–707

Silman, A.J. & Evans, S.J.W. (1981) Regional differences in survival from cancer. *Community Med., 3,* 291–297

Vincent, T.N., Satterfield, J.V. & Ackerman, L.V. (1965) Carcinoma of the lung in women. *Cancer, 18,* 559–570

Waterhouse, J., Muir, C.S., Shanmugaratnam, K. & Powell, J., eds (1982) *Cancer Incidence in Five Continents Vol. IV (IARC Scientific Publications No. 42)*, Lyon, International Agency for Research on Cancer

WHO/IARC Expert Committee on Cancer Statistics (1979) *Cancer Statistics (WHO Technical Report Series 632)*, Geneva, World Health Organization, pp. 24–26

Wynder, E.L., Bross, I.J., Cornfield, J. & O'Donnell, W.E. (1956) Lung cancer in women—a study of environmental factors. *New Engl. J. Med., 255,* 1111–1121

Young, J.L., Percy, C.L., Asire, A.J., Berg, J.W., Cusano, M.M., Lynn, B.A., Gloeckler, A., Horm, J.W., Lourie, W.I., Pollack, E.S. & Shambaugh, E.M. (1981) Cancer incidence and mortality in the United States, 1973–77. *Natl Cancer Inst. Monogr., 57,* 1–9

8. CANCER CARE PROGRAMMES: THE SWEDISH EXPERIENCE

T.R. MÖLLER

Southern Swedish Regional Tumour Registry, University Hospital, S-221 85 Lund, Sweden

SUMMARY

The present organization of oncological care in Sweden is described. In each health care region, there is an oncological centre with specific functions, including responsibility for running a regional tumour registry and for developing cancer care programmes for different types of tumours. The regional tumour registry is population-based, operates in close contact with the national central cancer registry and has, in addition to cancer registration, the key function of coordinating and evaluating cancer care programmes.

Cancer care programmes are defined as agreed means of managing patients with a specific disease which are effective throughout a region and at all levels within a health care system. They represent the right of all patients with cancer within a health care region to equal, standardized care, regardless of their domicile. Such standardized management also forms an excellent basis for performing clinical trials within different cancer care programmes.

Current cancer care programmes in Sweden are listed, together with some examples of patient accrual in trials within regional and national programmes.

INTRODUCTION

Cancer treatment is a multidisciplinary task. This realization has led to the establishment of specialized cancer hospitals and comprehensive cancer centres and their equivalent. Patients are usually referred to such institutions for treatment and generally also for follow-up.

Clinical trials are also an important tool in improving cancer treatment. Comparisons of different treatment regimes and evaluations of new drugs and techniques are usually done using a randomized design. Various criteria are used to decide on which type of patient is entered into a trial. Unless the cancer centre has a very well-defined catchment area, patient accrual may differ from one trial to another. Usually, several centres collaborate in a trial, in order to obtain the required number of patients in as short time as possible, although this process introduces selection bias.

The results of a trial may have a great impact on the management of a larger number of patients with that particular disease than those included in the trial. The justification for extrapolating the results to the entire population is usually not investigated. In certain areas, the effects of a treatment on a wider population have been monitored to some extent by following trends in survival from population-based registries. However, in such cases, many assumptions must be made regarding the actual management of patients. It would be of advantage, therefore, if trials could be carried out in a population-based set-up in order to evaluate not only survival but also other parameters (e.g., relapses, quality of life, use of resources, etc.).

The organization of cancer care in Sweden (which implements the main ideas proposed in *WHO EURO Report 70:* WHO Regional Office for Europe, 1982) makes it possible to carry out trials in this way. This paper reports some of these concepts and experiences.

ORGANIZATION OF CANCER CARE IN SWEDEN

In 1974, the National Board of Health and Welfare in Sweden proposed a new plan for organizing oncological care. Within each of the seven (presently reduced to six) health care regions in Sweden, an oncological centre was to be established (Fig. 1 and Table 1). An oncological centre comprises administrative and functional coordination of all resources for cancer care within the regional hospital. All radiotherapy with curative intent as well as more complex chemotherapy is delivered at the department of oncology of the hospital. That department is also given the responsibility for the functioning of the oncological centre. The tasks of the oncological centre are defined as follows:
— to set up and run a regional cancer registry, generate cancer statistics pertinent to the region, and standardize record keeping for cancer patients;
— to initiate cancer care programmes;
— to coordinate the resources for cancer care at the regional hospital and within the region as a whole;
— to give advice and information about oncological matters to members of the medical profession in the region;
— to provide for training and education of personnel of all categories;
— to promote theoretical and clinical research;
— to supervise the psychological and social aspects of oncology; and
— to give advice about mass screening activities and public information.

The main concept behind these recommendations was the right to equal care for all patients with cancer within the health care region, regardless of their domicile, and the optimal utilization of the combined treatment resources within the area. This concept is materialized into a cancer care programme for each specific type of tumour. The Oncological Centre of the Southern Health Care Region, in Lund, may be taken

Fig. 1. Map of Sweden, indicating the different health care regions

Table 1. Population and number of new cancer cases per health care region in Sweden, 1980

Region	Population (millions)	No. of cancers
Umeå	0.91	3 434
Uppsala-Örebro	1.89	7 782
Stockholm	1.59	6 497
Göteborg	1.52	6 132
Linköping	0.94	4 097
Lund	1.47	6 744
Total	8.32	34 686

as an example of the organization of an oncological centre (Fig. 2). Similar organizational structures have been implemented in most regions.

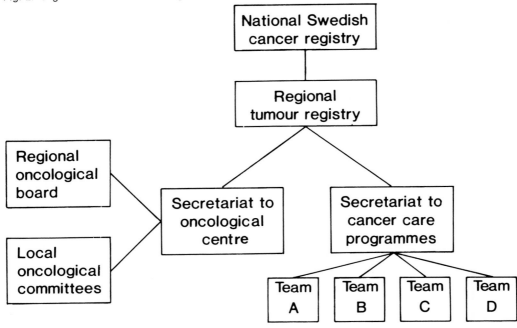

Fig. 2. Organization of the Oncological Centre of the Southern Swedish Health Care Region

The basic management of cancer patients is the responsibility of a number of teams, each constituting several specialists for a specific type of tumour, e.g., mammary carcinoma, ear-nose-and-throat tumours, orthopaedic tumours. Such teams represent the multidisciplinary approach to cancer management. Besides actual patient management, the team usually also forms the expert group on that particular disease and suggests and initiates cancer care programmes.

Local oncological committees are established within each hospital. Members of local committees form the regional oncological board, which makes all ultimate decisions on cancer care programmes and on other oncological issues for the region. A secretariat at the regional tumour registry serves these administrative organs.

Reports on new patients and follow-ups of old patients are submitted to the secretariat of each cancer care programme. Since the secretariat is attached to the regional tumour registry, the registry (the tasks of which are described below) thus has the responsibility for conducting the different activities.

REGIONAL TUMOUR REGISTRIES

Reporting of new cases of cancer has been compulsory in Sweden since 1958, the year in which the National Cancer Registry was started. Since then the Registry has served as a population-based cancer registry and publishes annual cancer incidence

Fig. 3. Starting dates of the regional tumour registries in Sweden

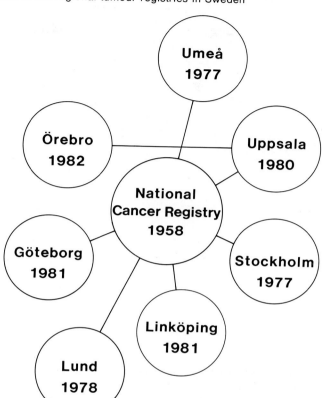

figures in a publication, *Cancer Incidence in Sweden* (National Board of Health & Welfare, 1983). The information available for each case is, however, relatively limited.

At the same time as oncological care was reorganized, as described above, the processing of cancer reports was switched from the central to regional registries. All management of primary data, requests for additional information, coding, data entry and quality assurance are thus the task of regional tumour registries. At the end of each calendar year, a data file is sent to the central registry, which has thus now become more of a coordinating institution, ensuring consistency of coding. Presentation of annual incidence figures is still the responsibility of the central cancer registry and, since it has now been relieved of the workload of such activities as coding, it can devote more resources to epidemiological research.

The regional tumour registries started at different points in time (Fig. 3), but now all health care regions possess a regional registry. The tasks of these registries are the following:

– to collect data on all new cases of cancer;

- to report to the central cancer registry;
- to present cancer statistics pertinent to the region;
- to take part in the design of therapeutic trials and protocols;
- to collect, evaluate and present data on cancer care programmes;
- to supervise the follow-up of cancer patients;
- to perform epidemiological studies; and
- to evaluate the use of resources for cancer treatment.

Their three main tasks are thus cancer registration, the supervision and evaluation of management of different types of cancer, and a wide range of epidemiological research. The registries are all population-based, and their computer files are updated regularly by matching to population registers and death certificate registers. This enables a close follow-up of all cancer patients within the region, regardless of whether they are entered into a trial or not.

CANCER CARE PROGRAMMES

Principles

The definition of a cancer care programme is an agreed means of referral, diagnosis, classification and staging, treatment and follow-up of *all* patients with a specific disease within a region. Such programmes therefore do not only specify the medical management of a specific disease, but also decide on the appropriate health care facilities needed for carrying out specific measures.

The setting-up of a cancer care programme is usually quite a complicated affair. A preliminary version is often drafted by the tumour team at the regional hospital and discussed with representatives of all disciplines involved in the care of that particular disease. Regional meetings are usually held before the final decision is taken. The final programme must be approved by the regional oncological board (for the Southern Swedish Health Care Region) or by equivalent bodies in other regions.

The cancer care programme thus forms the basis for standardized management of patients. The establishment of guidelines for the different diagnostic and therapeutic measures to be carried out means that differences in management between hospitals in the region are minimized. It also provides an excellent background for including a clinical trial in the cancer care programme, with continuing possibilities for supervising patient accrual into the trial and management of patients inside and outside the trial.

Regardless of whether or not a trial is included in the cancer care programme, the programme itself must be evaluated periodically with regard to, e.g., resources spent, the sensitivity and specificity of the diagnostic procedures used, treatment outcome in terms of overall and recurrence-free survival, and assessment of the subsequent quality of life of treated patients.

Secretariat

Coordination of a cancer care programme is usually carried out by a secretariat or secretarial function set up at the regional tumour registry. The specific tasks of this secretariat may be summarized as follows:
- to coordinate the project;
- to organize regional meetings;
- to perform randomization in trial(s);
- to collect data;
- to enter data into the computer;
- to assess the validity of data;
- to report to participating clinics; and
- to publish results.

The personnel required for the secretarial function is determined by the number of patients entered into a study, the routine for reporting and the amount of information to be collected and processed. For programmes in which there is extensive data collection, a total of 500 patients seems to be a reasonable number that could be handled by a full-time secretary (bearing in mind that these patients are followed for a considerable length of time); in programmes which involve less extensive data collection more than 1000 patients could easily be dealt with by one person.

A prerequisite for the management of such programmes is free and easy access to various computational and statistical facilities.

The role of the regional tumour registry

The main functions of the regional tumour registry in cancer care programmes may be summarized as follows:
- to register all new patients;
- to supervise the follow-up of patients;
- to perform statistical analysis to evaluate diagnostic and therapeutic procedures; and
- to perform special studies related to the programme.

In order to accomplish these tasks, an efficient computer system is essential. For each programme, a data base must be defined, data collection routines worked out and search and output programmes implemented; finally, various statistical programmes should be available to analyse the collected data. Since the structure of a cancer care programme is similar regardless of the type of tumour, the data base may be structured in a similar way for all programmes (Fig. 4).

Of the six regional tumour registries in Sweden, four now have computers of their own (three utilize the data base system MUMPS, and one uses MIMER); the two remaining registries use the same large university computer installation.

Present state of the art

The starting dates of the different oncological centres and tumour registries are shown on Figure 3. Regional variations in existing routines and traditions of

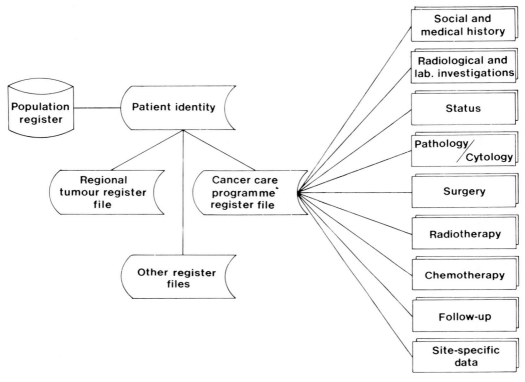

Fig. 4. Outline of data base structure for cancer care programmes and relations between different computer files

management and other factors, such as special interests in particular tumour types, have meant that different cancer care programmes have been developed.

The present programmes are listed in Table 2. For the most common tumour form, female breast cancer, programmes have been developed in almost all regions; the differences between some of these programmes are, however, minimal, and discussions are presently being held with a view to merging them into a national programme. National programmes have already been developed for some of the minor tumour groups. Programmes for other common tumours are implemented in the larger regions.

DISCUSSION

It has been estimated that about 30 different cancer care programmes can cover the entire spectrum of malignant diseases. However, since substantial effort and resources are required for a programme to be established and to be effective, the rate of implementation has been relatively slow—generally, about two programmes per year.

Table 2. Cancer care programmes in Sweden (as of 31 December 1983)

Lund	Linköping	Göteborg	Stockholm	Uppsala-Örebro	Umeå
Regional programmes					
Breast cancer Ovarian cancer I-II Prostatic cancer I-II Malignant glioma	Breast cancer	Myeloma	Breast cancer Myeloma Rectal cancer Bladder cancer Acute lymphocytic leukaemia	Breast cancer	Breast cancer
National programmes [a]					
Non-Hodgkin's lymphoma					
low grade	+ [b]	+	− [b]	−	+
high grade	+	+	+	+	+
Ovarian cancer III	+	+	+	+	+
Soft-tissue sarcoma	+	+	+	+	+
Osteogenic sarcoma	+	+	−	+	+
Ewing's sarcoma	+	+	+	+	+
Non-seminomatous testicular tumours	+	+	−	+	+
Solid tumours of childhood	+	+	+	+	+
+	+	+	Oral cavity	+	+
+	+	+	Malignant melanoma	+	+
+	+	+	+	Non-small-cell lung cancer	+
(+)	(+)	(+)	(+)	(+)	Stomach cancer

[a] The secretariat is set up within the region under which the name of the tumour type appears.
[b] +, participating in a programme; −, not participating

This is due mainly to lack of resources, particularly manpower (secretaries and computer programmers), but also to the reluctance of physicians to deal with many reporting forms. Thus, in certain current programmes patients will be followed up largely through population registers.

The existence of personal registration numbers in Sweden makes feasible matching with other registers, e.g., of hospital in-patients and health insurance schemes, to obtain further information, such as number of days spent in hospital or on sick leave, as an objective measure of quality of life or performance.

Matching of a cancer care programme file to the regional tumour registry file makes possible the study of patient referral patterns and variations in the selection of patients for the care programme. Thus, in a recent analysis of a trial within the care programme for mammary carcinoma stage II in southern Sweden, it was found that 81% of all potential patients had been entered into the study. Patients who were not entered fulfilled one (or more) of the ineligibility criteria, such as bilateral breast cancer, concurrent malignant or other serious disease, or non-radical operation (Möller et al., unpublished data). This group of patients will, however, be evaluated in the same way as the patients in trial.

In a study of ovarian carcinoma, 75% of all possible patients were referred to the regional department of gynaecological oncology for treatment. The median survival time of these patients was 32 months, whereas the median survival time of patients not referred was two months. This indicates that patient selection may have a profound influence on survival figures obtained from hospital-based patient material (Sigurdsson et al., 1983).

The two examples just cited are clinical trials carried out within regional cancer care programmes, which explains the relatively high accrual rate of patients. In a national study performed within a cancer care programme for non-Hodgkin's lymphomas, however, patient accrual varied considerably between the different participating regions, with a minimum patient input of less than 10% in one region and a maximum accrual of more than 50% in another. In addition, the proportion of patients in stage I and II varied between regions (Cavallin-Ståhl et al., 1984; Hagberg & Lindemalm, 1984; Möller & Mattsson, 1984). These differences reflect the degree of participation of departments outside the department of oncology in the respective regions and variations in diagnostic procedures, even when the agreed protocol is followed closely.

Such variations are often observed in multicentre trials, involving the participation of departments with widely varying catchment areas and patient management traditions. Even if clinical trials are evaluated according to commonly accepted principles, such as those established by Peto et al. (1976, 1977), some of the inherent problems such as selection bias would be lessened if the trial included patients from only one region all of whom were treated according to an agreed programme (Grage & Zelen, 1982).

More than 10 000 patients have now been reported to the existing cancer care programmes, and in the future a sufficient number of patients will be available for evaluation in most programmes. The results of such evaluations will, of course, form the basis for re-evaluation and possible revision of the programmes. The regional tumour registries together with their secretariats have the key functions in this process.

REFERENCES

Cavallin-Ståhl, E., Johnson, A. & Landberg, T. (1984) *A national cancer care program for non-Hodgkin's lymphoma in Sweden—Part I. Adjuvant chemotherapy after* radiotherapy for localized non-Hodgkin's lymphoma. In: *Proceedings, Second International Conference on Malignant Lymphoma, Lugano*

Grage, T.B. & Zelen, M. (1982) *The controlled randomized clinical trial in the evaluation of cancer treatment—the dilemma and alternative trial design.* In: *Evaluation of Methods of Treatment and Diagnostic Procedures in Cancer (UICC Technical Report Series Vol. 70)*, Geneva, International Union Against Cancer, pp. 23–46

Hagberg, H. & Lindemalm, C. (1984) *A national cancer care program for non-Hodgkin's lymphoma in Sweden—Part III. CHOP versus MEV for the treatment of non-Hodgkin's lymphoma with unfavourable histopathology.* In: *Proceedings, Second International Conference on Malignant Lymphoma, Lugano*

Möller, T. & Mattsson, W. (1984) *A national cancer care program for non-Hodgkin's lymphoma in Sweden—Part II. Prednimustine versus CVP in the treatment of non-Hodgkin's lymphoma with favourable histopathology.* In: *Proceedings, Second International Conference on Malignant Lymphoma, Lugano*

National Board of Health & Welfare (1983) *Cancer Incidence in Sweden 1980*, Stockholm, The Cancer Registry

Peto, R., Pike, M.C., Armitage, P., Breslow, N.E., Cox, D.R., Howard, S.V., Mantel, N., McPherson, K., Peto, J. & Smith, P.G. (1976) Design and analysis of randomized clinical trials requiring prolonged observation of each patient. I. Introduction and design. *Br. J. Cancer*, **34**, 585–612

Peto, R., Pike, M.C., Armitage, P., Breslow, N.E., Cox, D.R., Howard, S.V., Mantel, N., McPherson, K., Peto, J. & Smith, P.G. (1977) Design and analysis of randomized clinical trials requiring prolonged observation of each patient. II. Analysis and example. *Br. J. Cancer*, **35**, 1–39

Sigurdsson, K., Johnsson, J.-E. & Möller, T. (1983) Treatment of ovarian cancer in the Southern Swedish Health Care Region during the five-year period 1974–1978. *Ann. chir. gynaecol.*, **72**, 260–267

WHO Regional Office for Europe (1982) *Development of Cancer Centers and Community Cancer Control Programmes (EURO Report and Studies 70)*, Copenhagen

9. SERVICE ROLE OF THE HOSPITAL TUMOUR REGISTRY IN THE USA

C. ZIPPIN[1]

Cancer Research Institute, Department of Epidemiology and International Health and Department of Pathology, School of Medicine, University of California, San Francisco, CA 94143, USA

M. FEINGOLD

Oncology Program, St Vincent Medical Center, Los Angeles, CA 90057, USA

INTRODUCTION

More than 1000 hospital tumour registries exist in the USA, which were established primarily as components of hospital cancer programmes, within which they provide an important service. Many hospital registries contribute data to centralized programmes, and also participate in local, state or national cancer control efforts, including the evaluation of early detection programmes, epidemiological investigations and controlled clinical trials for evaluation of therapy. This paper reviews some primary aspects of hospital and centralized cancer registry activities in the USA, with emphasis on the service function of hospital registries, and discusses possible directions that these registry programmes may follow in the future.

CANCER REGISTRATION IN THE USA

Hospital registries

Hospital tumour registries or their equivalents have existed in the USA for many decades. Reports on cancer experience were issued in the late 1800s from institutions such as the American Oncologic Hospital in Philadelphia, the General Memorial Hospital in New York City, the Johns Hopkins Hospital in Baltimore, and Charity Hospital in New Orleans.

The modern impetus for the establishment of hospital cancer registries in the USA comes mainly from the work of the Commission on Cancer of the American College

---

[1] To whom requests for reprints should be addressed

of Surgeons. The Commission has surveyed hospital cancer programmes since the 1930s and has published periodically a list of institutions with programmes that meet the standards of the College for effecting a decrease in the morbidity and mortality of cancer patients. In order for its cancer programme to be approved, a hospital must make provision for educational facilities for its medical staff, for evaluation of quality of care and for monitoring its experience in the treatment of cancer patients.

In all these functions, the hospital cancer programme depends critically on a productive tumour registry. As early as 1956, the American College of Surgeons made approval contingent on the establishment of a cancer registry providing life-long follow-up of cancer patients. These requirements have been further refined, and now hospitals are categorized according to size and facilities, and the standards to be maintained by each category of institution are specified (Commission on Cancer, 1980).

In the past, financial support for oncology residencies or fellowships from some American health agencies was available only to institutions with cancer programmes approved by the American College of Surgeons. Such institutions also benefit from other forms of recognition and financial support.

At present, 1080 hospitals have cancer programmes approved by the American College of Surgeons, and these institutions treat 60% of all cancers diagnosed in the USA each year (Smart, 1983).

Central registries

Centralized registries bring together data from more than a single source. The American College of Surgeons has identified 39 states in the USA that have central cancer registry programmes, including both population-based and non-population-based activities. Some are incidence registries, such as the New York State Cancer Registry, which does not follow up patients regularly to assess survival. Others, such as the Connecticut Tumor Registry, collect both incidence and follow-up data. The Rocky Mountain Cancer Data System (Smart, 1983) includes both hospital-based and population-based programmes and concentrates on the clinical aspects of cancer, its management and survival results. Recently, another programme, the Centralized Cancer Patient Data System (CCPDS), has collected the data of about 20 comprehensive cancer centres located throughout the USA.

Continuous collation of inter-regional registry data was first undertaken in the USA in 1956 by the National Cancer Institute with the establishment of the End Results Program (Cutler & Latourette, 1959). Contributors to this programme were population-based and non-population-based central registries, as well as a number of large individual hospital registries located across the country. The End Results Program thus began at approximately the same time that hospital tumour registries were rapidly growing in number due to the requirement of the American College of Surgeons that hospital cancer programmes include an active tumour registry. In order to ensure comparability of data, the End Results Program helped to standardize the basic data set employed by most American hospital registries and to define the procedures to be followed.

The End Results Program was the forerunner of the present SEER (Surveillance, Epidemiology, and End Results) Program (Young et al., 1981) of the National Cancer Institute, a population-based activity that collects incidence and survival data from the states of Connecticut, New Jersey, Iowa, Utah, New Mexico and Hawaii, as well as from the Detroit, Atlanta, Seattle and San Francisco metropolitan areas, and the Commonwealth of Puerto Rico. The standardization of data sets, the definitions and the quality control standards recommended by the SEER Program have had a continuing effect on improving hospital registries throughout the USA.

Training for tumour registrars

Increasing emphasis has been placed in recent years on the education of tumour registrars. In 1961, a regularly scheduled training programme (Zippin, 1978) was initiated and continues to be offered by the Cancer Research Institute, University of California in San Francisco. Other training workshops were organized by the American College of Surgeons. More recently, a major contributor to the continuing education of tumour registrars in the USA has been the National Tumor Registrars Association, a professional organization for persons in registry work, which was established in the early 1970s. This Association administers national examinations leading to the title of Certified Tumor Registrar. Local and state registrar associations have been established in all 50 states, providing a firmer basis for the professional development of registry personnel.

An additional educational aid is a series of self-instruction manuals for tumour registrars published by the SEER Program. Six volumes (Shambaugh, 1976a,b, 1977a,b,c, 1980) have been published on various aspects of registry work, including the objectives and functions of tumour registry, cancer characteristics and selection of cases, tumour registrar vocabulary, human anatomy as related to tumour formation, abstracting a medical record, and classification for extent of disease. A seventh volume on statistical methods for tumour registrars is in preparation.

Quality control

Activities as widespread and as costly as tumour registration require constant quality assurance. Increasing awareness of potential problems related to data quality has resulted in major efforts both to quantify and reduce the frequency of unreliable data. Major efforts in this direction have been made by the SEER Program and the Centralized Cancer Patient Data System of the Comprehensive Cancer Centers Program (Feigl et al., 1982) and by the National Tumor Registrars Association.

SERVICE FUNCTIONS OF THE HOSPITAL REGISTRY

The focus of hospital cancer programmes in the USA has broadened in recent decades to include an array of clinical and educational activities; in many of these the tumour registry plays an important role. The data collection, retrieval and analysis activities of tumour registries have long been accepted as essential by physicians and epidemiologists concerned with assessing cancer incidence, treatment and end results; however, the extent to which registries can assist hospital planners, administrators and the community at large are less widely appreciated. This is described below.

Record-keeping function

Most hospital tumour registries in the USA now perform standard record-keeping tasks in a fundamentally similar fashion. These activities include:
– identification of all new patients with reportable diseases, possibly including some benign tumours of interest;
– collection from medical records of core items of information on each reportable case, such as date of diagnosis, primary site, histological type, stage/extent of disease, initial course of treatment and patient identifying information;
– coding of some or most of these core items of information to facilitate retrieval, comparison and analysis of data;
– annual follow-up of patients to determine vital status and disease status;
– reporting of registry data to hospital medical and administrative staff on at least an annual basis, including survival information for selected sites of cancer;
– participation in studies to evaluate patient care, which assess both patterns of care as well as long-term outcomes.

The maintenance of extensive records for determining end results may not be practical for individual physicians who treat and follow cancer patients on either an inpatient or outpatient basis. Since almost all cancer patients are diagnosed and/or treated in hospitals, hospital tumour registries that utilize data from pathology, haematology, radiology, surgery, medical records and other departments provide a relatively complete data system for reportable cases. As data retrieval by registries increasingly becomes automated, individual physicians can receive comprehensive information about their own experience in treating cancer patients.

Some hospital registries maintain even more extensive data bases, which may include, for example, information on subsequent as well as initial course of therapy. This can be useful in following the natural history of the disease, as well as providing information on patterns of care. Some registries that are affiliated with institutions where many cases are seen can use computer systems to generate appointments for follow-up visits for patients, eliminating the need for physicians to carry out an essentially clerical function.

Case management

Tumour registry data can provide guidance to physicians in deciding on treatment and follow-up: as the repository of an institution's historical cancer experience, the registry can assist in identifying groups of patients who have presented in the past with particular characteristics, describe how they have been managed, and offer information on the results of treatment.

Every hospital in the USA with an approved cancer programme is required to provide multidisciplinary consultative services. Most institutions choose to organize this service in the form of a tumour board, also called a 'cancer conference', which is composed of physicians representing all disciplines involved in cancer care. Tumour boards may meet regularly and have a prospective, case-oriented approach, rather than a retrospective, didactic one. Although the ultimate decision about appropriate treatment rests with the physician and patient, a tumour board can recommend one or more courses of action that would result in the most positive outcome for the patient. At a tumour board conference, tumour registry data may be used to summarize an institution's past experience with a particular disease, representing a factual, rather than an anecdotal, approach to case management.

Information presented at a tumour board meeting may also ultimately affect patterns of diagnosis and care in a community. For example, comparison with state or national data of the distribution of stage of disease at diagnosis for a particular site of cancer in a given community has encouraged the development of cancer screening programmes and improved medical education in many communities in order to reduce the proportion of patients diagnosed with advanced disease (Hanlon, 1983).

Hospital statistics

Much of the data collected by a registry can aid hospital planners and administrators in analysing the impact of a cancer programme—its strengths and weaknesses and the characteristics of the area and population served—and in planning future resources and activities.

Review of the annual report of a tumour registry, paying particular attention to the distribution of new cases by primary site and stage within site, may be quite useful. A particular hospital may see a fairly standard distribution of cancers at the common primary sites or may, for some reason, treat unusually large numbers of patients with a cancer of a particular site or organ system. For example, if a hospital planner finds that 15% of cancer cases are bone/soft-tissue sarcomas, frequently requiring limb amputations and other radical surgery, he may wish to develop a strong rehabilitation programme or to formalize a referral relationship with another institution that has those resources. An unusual distribution of cancers seen in an institution may also reflect the presence of expertise in specialized fields, such as paediatric oncology or brain tumour research.

Patient characteristics, such as age, race, sex, stage of disease at diagnosis and payment source, may vary from one community to another, and may suggest directions for cancer control efforts. The presence of an elderly population at high risk of developing cancer may suggest a need for intensive, site-specific education and screening programmes. Since incidence figures for many cancers vary by race, a community with a large ethnic minority at high risk of one cancer may require special programmes concentrating upon high-risk groups. Comparison with similar data from neighbouring hospitals, from state or regional registries, or from the National Cancer Institute may reveal gaps in a cancer programme or suggest re-allocation of resources as patient subgroups grow or shrink in size.

Hospital planners and administrators, as well as physicians, can use registry data to monitor high-risk groups exposed to occupational or industrial carcinogens. While population-based central registries may be better equipped to collect and analyse the large amount of data required to assess actual risk, hospital data may also show trends of importance and point to the need for community-wide screening programmes. For example, in many hospital and central registries in the coastal areas of the USA, sharp increases in the number of mesotheliomas being diagnosed were noticed in the last decade (Selikoff, 1976). A number of studies revealed that many of the patients had similar occupational histories: employment in the ship-building industry and heavy exposure to asbestos during the 1940s and subsequently. This finding led to the establishment of education and screening programmes throughout the region, many of which were initiated by community hospitals.

Most hospital cancer programmes in the USA are run by specialists in medical oncology, radiology and surgery, to whom cancer patients are referred by internists and by general and family practitioners. As hospitals become increasingly competitive in their efforts to maintain adequate revenue to continue their programmes, administrators must know from where their patients come, both in terms of residence and referring physician. Studies of this kind can be based on registry data.

Studies of place of residence of cancer patients help to determine the service area for a hospital's oncology programme. When such studies are linked with information collected by the US census, hospitals can gain valuable information about the characteristics of the population in their geographic areas in order to be able to provide relevant services.

Knowledge of patterns of primary referral by physicians to oncological specialists can assist administrators in strengthening relationships between hospitals and referring physicians. By involving such physicians directly in the cancer programme, by inviting them to educational courses at the hospital, and by assuring that they are kept informed of their patients' progress as treatment progresses, these important relationships can be maintained.

After-care and follow-up

Identifying cancer patients for annual follow-up examinations is difficult if not impossible for most private physicians, particularly if the size of their practice is increasing and since many patients now survive longer. The follow-up system of a hospital registry ensures that every patient not known to be dead is followed within 12 months of the date he or she was last seen by a physician or known to be alive.

Typically, a registrar sends a letter to the patient's physician requesting information on the status of the patient when last examined. If the patient was seen in the past year, that information is entered on the follow-up letter by the physician or office nurse, and returned to the registry. However, if the patient did not return as requested, the physician contacts the patient to urge a visit, to establish whether there has been an early recurrence, a new primary tumour or that the patient is free of disease.

The presence of a tumour registry in a community would thus appear to have a positive impact on public health: its follow-up function benefits both the physician and the patient, who receives an annual reminder to return for a medical examination. Early detection of a recurrence or a subsequent primary tumour can lead to timely treatment and the expectation of a more favourable prognosis. Long-term benefits include the prospect of better management of future cancer patients in that community as the medical establishment gains expertise in oncology, and technical capabilities for diagnosis and treatment expand as a result of the focus on cancer care which the registry provides.

Professional education and audit

The American College of Surgeons requires that each approved hospital cancer programme utilize registry data in the continuous education of staff physicians and in assessing the quality of cancer care provided in the institution. Annual reports, patient origin, referral data and other kinds of information can provide the basis for such activities.

Audit, or assessment of the quality of care, is closely associated with professional education, since the goal of auditing is to provide feedback to practitioners. Audits may compare actual care with a standard intended to represent the most appropriate care for a given diagnosis, and explanations of differences between the standard of care and actual practice are requested and reviewed. Educational programmes are frequently designed to ensure that practitioners in a given community are aware of the newest approaches to cancer diagnosis and treatment.

Audits can reveal changes in patterns of care over time. An audit of breast cancer treatment carried out by the American College of Surgeons revealed that 50% of patients received radical mastectomies in 1972 and only 3% were so treated in 1981 (Hanlon, 1983). The predominant form of surgery in 1981 was modified radical mastectomy, and end results data, disseminated in part through hospital tumour registries, indicated no reduction in survival related to the latter procedure. Another audit carried out by the American College of Surgeons in the 1970s showed a rapid increase in the incidence of benign liver tumours in young women. Hospital registries were asked to identify recent cases and to complete detailed abstracts, and a statistically significant link with use of birth control pills was quickly established (Baum *et al.*, 1973).

Physicians who prepare articles for publication in journals or who speak at medical conferences on topics related to oncology may make frequent use of hospital registries as a primary source of data. Many registries with computer facilities and statistical staff collaborate with researchers in study design, analysis of data and dissemination of results. Such support might be prohibitively expensive if purchased elsewhere.

IMPACT OF REGISTRY ACTIVITIES ON EARLY DIAGNOSIS, MANAGEMENT AND SURVIVAL RATES

It is difficult to assess directly the effect of cancer registries on diagnosis, management and survival. However, it is fair to say that the availability of registry information has made it possible to compare cancer experience in different institutions, in different geographical areas and at different points in time. Comparisons can be made of distribution of primary site, histological type, method of diagnosis, extent of disease, treatment and survival outcome in an increasingly standardized fashion.

End Results Program and SEER Program

The End Results Program of the National Cancer Institute, which collected data from 14 registries, was established just as large-scale testing of cancer chemotherapeutic agents began in the USA. One of the original goals of the Program was thus to provide baseline data on cancer diagnosis, management and survival to be compared with data obtained following the introduction of chemotherapy.

Reports from this group were presented at a national cancer conference in sessions devoted to cancer management and end results (Hickey & Showalter, 1960; Latourette, 1960; Mersheimer & Ederer, 1960), and some of the findings were hotly debated by proponents of one or another method of cancer management. Reports on survival were widely disseminated and the findings used in numerous papers by other clinicians. Considerable attention therefore was given in the USA to this registry programme, which represented the best available summary of national experience in cancer diagnosis and management. Individual hospitals were encouraged to compare their own results with the national findings.

Data on survival of patients with cancers of the colon and rectum from hospital tumour registries that contribute to studies conducted by the American College of Surgeons have shown considerably greater variability from state to state than would have been expected if only sampling variability were operating (Zippin & Zippin, 1983). Even for localized (stage I) disease, the standard deviation of the survival rates observed in different areas for patients with colon cancer was more than twice as great as expected. For localized cancers of the rectum, the observed standard deviation exceeded by approximately two-thirds that which would have been expected if only random variability were operating. For localized cancer of the female breast, the observed variability exceeded that expected by approximately 35% (Zippin & Zippin, personal communication). It is to be hoped that these observations will stimulate studies to identify the reasons for these large differences and that subsequent changes in practice can lead to more uniformly satisfactory end results in the future.

It should be noted that Enstrom and Austin (1977) have cautioned that in comparing survival results from one time period to another, one must take into account possible changes in the characteristics of patients and of the disease or other factors that could contribute to differences in survival over time.

Evaluation of therapy

Conventional hospital tumour registries cannot make definitive evaluations of therapy in the absence of controlled clinical trials. They can, however, review results in different treatment groups and, when differences are noted, can contribute to determining what trials should be carried out. Tumour registry data from the USA and elsewhere were used in this way prior to the studies carried out in the USA in the 1970s to compare various forms of breast cancer treatment (Fisher *et al.*, 1980).

FUTURE DIRECTIONS

As data systems, cancer registries lend themselves to the automated methods that are becoming more widespread throughout the world. The Commission on Cancer of the American College of Surgeons (1983) has developed a software package, CANSUR, to facilitate the computerization of hospital registries. Increasing automation has implications in terms of the number and qualifications of registry personnel; additional adjustments will be necessary with regard to control of data quality, maintenance of confidentiality, registry space requirements, as well as operational aspects of the programmes.

Developments are also under way that will lead toward computer linkages between individual registries and central registries to which hospitals contribute data. Potential benefits of such a development include possible sharing of follow-up information about patients seen at more than one hospital, improved data quality due to the use of advanced computer editing systems, and access by hospital registries to a larger pool of data.

With increasing automation, it may be possible to include more details of treatment plans, dosages, etc, than currently found in most registry systems. This might assist potential users of the data in their assessment of past and current practices. The danger of such extension, however, is that a registry data set will eventually resemble a medical record. It is well known that no matter how much information is entered into a registry there will always be situations in which details desired by a clinician or researcher have not been included: it is impossible to know in advance in what directions cancer research might turn in the future and to anticipate what information will be needed. It is therefore important to remember that, apart from recording information of continuing value, the registry should be thought of as a device for identifying patient groups with specific characteristics and that more detailed information needed for specific studies may be found from the basic medical records of the patients.

Just as with tumour registries, attempts are being made in hospitals in the USA to computerize all or parts of medical records. As this proceeds, efforts will be made to integrate the registry system into the medical record department system. In this event, caution must be exercised to assure that personnel responsible for the cancer data portion of the system know how to abstract and code these data, and to analyse and interpret them.

As the costs of hospital care in the USA are rising rapidly, programmes of questionable value are being discontinued. This move represents a challenge for registries, some of which already have been threatened with closure: if a registry is not used adequately, it is indeed hard to justify its continuance. With the application of computer technology to registries, they must become more efficient and more widely used.

It is encouraging that the number of hospital cancer programmes approved by the American College of Surgeons increased from approximately 700 in 1979 to over 1000 in 1983. In addition, the number of centralized registries, primarily population-based, has also increased. It is to be hoped that the effort involved in this considerable undertaking will be rewarded by meaningful contributions from these programmes in the fight against cancer.

REFERENCES

Baum, J.K., Holtz, F., Bookstein, J.J. & Klein, E.W. (1973) Possible association between benign hepatomas and oral contraceptives. *Lancet, ii,* 926–929

Commission on Cancer (1980) *Cancer Program Manual,* Chicago, American College of Surgeons

Commission on Cancer, American College of Surgeons (1983) CANSUR update. *Field Liaison Newsletter,* Chicago, pp. 9–10

Cutler, S.J. & Latourette, H.B. (1959) A national cooperative program for the evaluation of end-results in cancer. *J. natl Cancer Inst.,* **22,** 633–646

Enstrom, J.E. & Austin, D.F. (1977) Interpreting cancer survival rates. *Science,* **195,** 847–851

Feigl, P., Polissar, L., Lane, W.W. & Guinee, W. (1982) Reliability of basic cancer patient data. *Stat. Med.,* **1,** 191–204

Fisher, B., Redmond, C., Fisher, E.R. and participating NSABP investigators (1980) The contribution of recent NSABP clinical trials of primary breast cancer therapy to an understanding of tumor biology—an overview of findings. *Cancer,* **46,** 1009–1025

Hanlon, C.R. (1983) Director's memo. *Bull. Am. Coll. Surg.,* **68**

Hickey, R.C. & Showalter, M. (1960) *End-results: Cancer of the breast—36,005 patients.* In: *Proceedings of the Fourth National Cancer Conference,* Philadelphia, J.B. Lippincott, pp. 251–262

Latourette, H.B. (1960) *End-results evaluation program data on cancer of the female genital tract.* In: *Proceedings of the Fourth National Cancer Conference,* Philadelphia, J.B. Lippincott, pp. 379–389

Mersheimer, W.L. & Ederer, F. (1960) *End-results evaluation of cancer of the lung and bronchus.* In: *Proceedings of the Fourth National Cancer Conference,* Philadelphia, J.B. Lippincott, pp. 319–333

Selikoff, I.J. (1976) Lung cancer and mesothelioma during prospective surveillance of 1249 asbestos insulation workers, 1963–1974. *Ann. N.Y. Acad. Sci.,* **271,** 448–456

Shambaugh, E.M., ed. (1976a) *Self-Instructional Manual for Tumor Registrars,* Book 2, *Cancer Characteristics and Selection of Cases (NIH Publication No. 76–993),* US Department of Health and Human Services, Public Health Service, National Institutes of Health

Shambaugh, E.M., ed. (1976b) *Self-Instructional Manual for Tumor Registrars,* Book 3, *Tumor Registrar Vocabulary: The Composition of Medical Terms (NIH Publication No. 76–1078),* US Department of Health and Human Services, Public Health Service, National Institutes of Health

Shambaugh, E.M., ed. (1977a) *Self-Instructional Manual for Tumor Registrars,* Book 1, *The Objectives and Functions of a Tumor Registry (NIH Publication No. 77–917),* US Department of Health and Human Services, Public Health Service, National Institutes of Health

Shambaugh, E.M., ed. (1977b) *Self-Instructional Manual for Tumor Registrars*, Book 5, *Abstracting a Medical Record: Patient Identification, History and Examinations (NIH Publication No. 77–1263)*, US Department of Health and Human Services, Public Health Service, National Institutes of Health

Shambaugh, E.M., ed. (1977c) *Self-Instructional Manual for Tumor Registrars*, Book 6, *Classification for Extent of Disease (NIH Publication No. 77–1448)*, US Department of Health and Human Services, Public Health Service, National Institutes of Health

Shambaugh, E.M., ed. (1980) *Self-Instructional Manual for Tumor Registrars*, Book 4, *Human Anatomy as Related to Tumor Formation (NIH Publication No. 80–2161)*, US Department of Health and Human Services, Public Health Service, National Institutes of Health

Smart, C.R. (1983) The potential for progress in cancer management today. *Bull. Am. Coll. Surg.*, **68**, 12–15

Young, J.L., Jr, Percy, C.L. & Asire, A.J., eds (1981) Surveillance, Epidemiology, and End Results: Incidence and Mortality Data, 1973–77. *Natl Cancer Inst. Monogr.*, **57**

Zippin, C. (1978) *Training for staff of cancer registration programs.* In: Nieburgs, H.E., ed., *Prevention and Detection of Cancer*, Part 1, *Prevention*, Vol. 2, *Etiology; Prevention Methods*, New York, Marcel Dekker, pp. 2165–2172

Zippin, C. & Zippin, D. (1983) *Survival from cancer of the colon and rectum in the United States: Comparing results from different sources.* Presented at Annual Meeting, International Association of Cancer Registries, Heidelberg, Federal Republic of Germany, September 1983

10. SECOND CANCERS AS A RESULT OF CANCER TREATMENT

N.E. DAY & G. ENGHOLM
International Agency for Research on Cancer, Lyon, France

INTRODUCTION

Cancer registries, as a rule, collect little concomitant information on registered patients that can be of use for analytical studies. Studies based on cancer registry material tend to yield general and rather anodyne conclusions, such as, for example, that parous women are at greater risk of cervical cancer, or at lower risk of ovarian cancer, than nulliparous women. The detailed information on reproductive history that would be needed, in this example, to allow more penetrating conclusions would not normally be available from the data sources routinely used.

There are areas, however, where routinely collected information can be particularly illuminating. One such field is the evaluation of changes in survival for specific cancers. Clinical trials indicate the benefits that can be obtained from alternative treatments, as do series from cancer centres or specialized clinics, but determination of whether improvements in treatment have filtered through to the general medical services requires population-based data. Some cancer registries—those of Norway (Cancer Registry of Norway, 1980) and Finland (Hakulinen *et al.*, 1981), for example —have been especially valuable sources of information in this respect, showing quite clearly where advances in treatment have had a major impact. A second area, less developed than survival analysis, is the study of second primary cancers. In both instances, a cancer registry can be a valuable source of material because the information required is an integral part of that which is routinely collected. Most registries match their records to national death records; many registries follow all individuals they register for life, and would expect to register subsequent malignancies. Study of second primary cancers, in particular investigation of whether there is an excess risk of cancer type B following cancer type A, has been used primarily as a basis for speculation on common pathways in the etiologies of the two tumours. Recently, however, some such studies have concentrated on a more precise target—the role that treatment for the initial cancer may have played in augmenting risk for a specific second cancer. Details of treatment are not, admittedly, very extensive in many registries, but the two examples that are examined in detail below

Table 1. Observed and expected cases of haematological cancer by time since diagnosis of cervical cancer for radiotherapy patients

Specification	Time since diagnosis (years)			
	0–3	4–8	9+	Total
No. starting	20 250	17 291	10 432	28 490[a]
Woman-years of observation[b]	49 495	49 367	35 578	134 440
Leukaemia:				
Observed	4	7	2	13
Expected	5.0	5.4	5.1	15.5
Observed/Expected	0.8	1.3	0.4	0.8
Lymphosarcoma, Hodgkin's disease, other reticuloses, multiple myeloma:				
Observed	1	7	7	15
Expected	7.1	7.5	6.9	21.4
Observed/Expected	0.1	0.9	1.0	0.7

[a] Because of the retrospective-prospective nature of the study design, many patients did not contribute years of observation during the years immediately after diagnosis of cervical cancer, i.e., not all 28 492 contributed woman-years during the 0–3 or 4–8 time periods.
[b] Enrollment to date of last known status

illustrate the utility of even fairly rudimentary information, especially if supplemented by data from additional available sources.

SECOND CANCERS FOLLOWING CANCER OF THE CERVIX

Radiation is widely used as part of primary treatment for cancer of the cervix. In Sweden, for example, nearly 100% of cases would receive radiation therapy (Malker & Pettersson, 1983). Furthermore, treatment regimens are largely standardized and well described in hospital records, enabling the radiophysicist to calculate with considerable accuracy the radiation dose received at other organs of the body, by means of phantom models and Monte Carlo techniques (Stovall, 1983). These calculations are particularly accurate for intracavitary radium, until recently the treatment of choice, especially for early-stage disease. Provided sufficient numbers of women are followed for a sufficient length of time, excess risk of second cancers at sites other than the cervix can be related quantitatively to the radiation dose received.

It is interesting to compare two approaches that have been used to study second primary tumours in cervical cancer patients. The first study, aimed primarily at assessing the risk of leukaemia as a consequence of the radiation, was clinically based (Hutchison, 1968). Many of the major treatment centres in Europe and the USA collaborated in this study, begun in the late 1950s. Each centre had access to extensive details of treatment on each patient, and each patient was followed intensively, clinically, and a blood sample was taken every six months. The results of ten years of follow-up have been published (Boice & Hutchison, 1980), and are shown in Table 1. If the dose-response for leukaemia were linear throughout the entire range of radiation doses, one would have expected several hundred cases of excess leukaemia on the basis of the experience of the Japanese atomic bomb survivors. If anything, however, a slight deficit was observed, although some excess, not approaching statistical significance, was seen four to eight years after treatment. Attempts have been made to extend the follow-up of this cohort beyond ten years, mainly to examine the risk of malignancies other than leukaemia. In some centres, no procedure exists for following up of individuals no longer under clinical surveillance, which often ceases ten years after treatment if no relapse has occurred. Follow-up on an ad-hoc basis is both laborious and costly. In other centres, no data are available on population incidence rates for the calculation of expected numbers. It is clear that the main value of extension of the follow-up period for this study will be to provide a basis for case-control studies, and that reliable information on absolute excess risk will be difficult to obtain.

The second approach for studying second primary cancers in patients with cervical cancer was through cancer registries, and was designed to investigate the excess risk for all second malignancies, not just leukaemia. Doubts were raised initially by clinicians on the periphery of the study as to its rigour and credibility: details about treatment in a cancer registry are sparse; it was considered that the follow-up, which depends on routine registration of second primaries, would lack depth compared to that in the earlier study in which blood samples were taken regularly; and the criteria for accepting a lesion as a second primary cancer vary between registries. The study was nonetheless carried out. Cancer registries were contacted which had been in operation for 20 years or more and which at least attempted lifetime follow-up of all registered cases. Most responded enthusiastically, and most of these could provide data of sufficient quality to merit inclusion in the study. The participating registries are shown in Table 2, together with the number of patients enrolled into the study. The size of the cohort is considerably greater than that of the earlier study (Table 1). Table 3 gives a summary of the observed and expected numbers of second primary cancers, by site, excluding the first year of follow-up to avoid problems associated with synchronous tumours.

The results for leukaemia are given in greater detail in Table 4, and merit some discussion, particularly in comparison with the results of Table 1. Table 4 gives a clear picture of the leukaemogenic effect of radiation treatment for cervical cancer. A peak of risk for myeloid and acute leukaemia, some two- to threefold, is seen in the first five years after irradiation, followed by a decline in the excess. No excess risk for chronic lymphocytic leukaemia is seen. The lack of excess in the earlier, clinic-based study is simply a reflection of inadequate numbers. A detailed, central review of the available haematological material will sharpen the different leukaemia diagnoses in the cancer registry study, but is unlikely to change the overall picture. Details of the dose of radiation given to active bone marrow are being obtained retrospectively for the leukaemia cases and a series of matched controls. Equivalent haematological and treatment information to that obtained in the clinic-based study is thus being acquired in the cancer registry study, on many fewer individuals but giving much more information. In brief, the earlier study put its initial emphasis on obtaining clinical information of high quality on the entire cohort, resulting in inadequate numbers to provide an informative conclusion. The later study, by contrast, initially spread the

Table 2. Numbers of women by registry, stage of cervical cancer and treatment status

Registry	Invasive cervical cancer		In-situ cancer
	Radiotherapy	No radiotherapy	
Canada:			
Alberta	2 096	411	—
British Columbia	2 232	81	230
Manitoba	1 337	1 035	4 832
New Brunswick	1 449	—	—
Nova Scotia	695	84	—
Ontario	7 322	717	—
Saskatchewan	1 277	232	—
Denmark	20 024	5 127	15 367
Finland	7 285	1 206	3 290
Norway	5 282	724	2 130
Sweden	14 760	—	43 971
United Kingdom:			
Birmingham	3 808	819	2 034
South Thames	6 110	1 217	3 774
USA:			
Connecticut	5 997	1 130	7 260
Yugoslavia:			
Slovenia	2 942	1 390	1 114
Total	82 616	14 173	84 002

net widely enough to generate a cohort of adequate size; the clinical information on each cohort member was rather sparse, but the necessary clinical details could be obtained retrospectively on relevant individuals.

The value of cancer registries as a basis for this type of study is well illustrated by the above example. It is not the purpose of this paper to reiterate the overall conclusions reached by the study (Day et al., 1983), but, even without supplementary information on dose of radiation, using only cancer registry records, results of great interest have been obtained concerning cancer of the breast (Table 5). The deficit of breast cancer among irradiated women under age 40 increases steadily as time elapses, demonstrating the central role of ovarian-dependent hormones in development of the disease. The excesses of multiple myeloma and of rectal cancer, and the negative results obtained with regard to cancers of the stomach and colon are striking (Table 6).

Even though information on confounding variables is not normally available in cancer registry records, information from other sources can be used to assist interpretation. Thus, among cervical cancer patients not treated by radiotherapy, the deficit of breast cancer is about 5% (Day et al., 1983, p. 174). Women with cancer

Table 3. Observed (O) and expected (E) numbers[a] of second primary cancers by stage of cervical cancer and treatment status

Second primary cancer (ICD7)	Invasive cervical cancer						In-situ cancer		
	Radiotherapy			No radiotherapy					
	O	E	O/E	O	E	O/E	O	E	O/E
Buccal cavity and nasopharynx (140–148)	60	46.61	1.3*	8	5.30	1.5	29	16.93	1.7**
Oesophagus (150)	40	27.32	1.5*	3	2.93	1.0	3	5.55	0.5
Stomach (151)	200	210.37	1.0	23	26.11	0.9	42	45.49	0.9
Small intestine (152)	21	9.46	2.2**	4	0.94	4.3*	3	3.70	0.8
Colon (153)	314	301.53	1.0	42	38.46	1.1	85	84.09	1.0
Rectum (154)	197	157.38	1.3**	29	21.78	1.3	48	43.00	1.1
Liver (155.0)	19	19.88	1.0	2	2.28	0.9	6	5.45	1.1
Gallbladder (155.1)	45	55.69	0.8	7	7.22	1.0	10	2.25	0.8
Pancreas (157)	120	95.88	1.3*	10	11.66	0.9	34	24.82	1.4
Nose (160)	13	7.12	1.8	2	0.72	2.8	6	2.21	2.7
Larynx (161)	16	7.21	2.2	4	1.10	3.6	5	3.33	1.5
Lung (162–163)	491	134.94	3.6***	60	20.96	2.2***	106	48.91	2.2***
Breast (170)	569	804.41	0.7***	118	124.91	0.9	409	426.59	1.0
Corpus uteri (172)	126	209.13	0.6***	2	31.44	0.1***	48	91.83	0.5***
Other uterus (173)	21	18.11	1.2	0	1.72	0.0	6	11.52	0.5
Ovary (175)	136	198.31	0.7***	16	30.95	0.5**	91	102.25	0.9
Other genital (176)	101	43.00	2.4***	17	5.94	2.9***	95	12.44	7.6***
Kidney (180)	69	66.50	1.0	18	8.41	2.1**	27	25.34	1.1
Bladder (181)	194	73.74	2.6***	16	9.26	1.7*	31	20.49	1.5*
Melanoma (190)	36	47.08	0.8	8	7.99	1.0	44	39.69	1.1
Other skin (191)	206	219.76	0.9	40	34.48	1.2	43	49.49	0.9
Eye (192)	7	8.47	0.8	2	1.04	1.9	4	3.84	1.0
Brain (193)	51	67.54	0.8*	16	10.34	1.6	39	42.33	0.9
Thyroid (194)	36	31.47	1.1	1	4.23	0.2	19	22.07	0.9
Bone (196)	11	5.72	1.9	0	0.69	0.0	2	3.08	0.7
Connective tissue (197)	27	14.57	1.9***	5	1.85	2.7	6	8.39	0.7
Lymphoma (200,202)	61	53.77	1.1	8	6.99	1.1	28	20.71	1.4
Hodgkin's disease (201)	14	16.87	0.8	3	2.46	1.2	14	11.21	1.3
Multiple myeloma (203)	33	33.92	1.0	1	4.02	0.3	12	10.56	1.1
All leukaemia (204)	77	65.83	1.2	13	8.76	1.5	39	24.95	1.6*
Chronic and unspecified lymphatic leukaemia (204.0)	18	22.64	0.8	6	3.04	2.0	7	6.07	1.2
Myeloid and acute leukaemia (204.1–4)	58	40.94	1.4*	6	5.41	1.1	31	18.25	1.7*
Close and intermediate sites[b]	1 601	1 479.27	1.1***	191	198.71	1.0	534	494.14	1.1
Close and intermediate sites excluding genital cancers (172–176)	1 217	1 010.72	1.2***	156	128.66	1.2*	294	276.10	1.1
Total (all sites except cervix)	3 324	3 062.54	1.1***	479	435.40	1.1*	1 346	1 238.05	1.1***

[a] Numbers exclude the first year of observation.
[b] Stomach (151), small intestine (152), colon (153), rectum (154), liver (155.0), gallbladder (155.1), pancreas (157), corpus uteri (172), other uterus (173), ovary (175), other genital (176), kidney (180), bladder (181), bone (196) and connective tissue (197).
* $0.01 < p < 0.05$; ** $0.001 < p < 0.01$; *** $p < 0.001$

Table 4. Leukaemia following irradiation for cancer of the cervix

	Time since irradiation (years)					Total
	<1	1–4	5–9	10–14	15+	
Chronic lymphocytic leukaemia						
Observed	3	4	5	2	7	18
Expected	2.35	6.60	6.26	4.63	5.15	22.64
Observed/Expected	1.28	0.61	0.80	0.43	1.36	0.80
Acute and myeloid leukaemia						
Observed	3	28	17	10	3	58
Expected	4.43	12.61	11.61	8.26	8.46	40.94
Observed/Expected	0.68	2.22***	1.46	1.21	0.35	1.42**

** $p<0.01$; *** $p<0.001$

Table 5. Breast cancer following cancer of the cervix (including in-situ carcinoma)

Age at diagnosis of cervical cancer and treatment		Time since diagnosis of cervical cancer (years)						Total[a]
		<1	1–4	5–9	10–14	15–19	20+	
<40 years								
Radiation treatment	Observed	2	9	9	15	5	3	41
	Observed/Expected	0.50	0.48	0.30	0.49	0.25	0.18	0.35
No radiation treatment	Observed	6	44	63	44	11	1	163
	Observed/Expected	0.55	0.80	0.95	1.31	0.82	0.21	0.94
≥40 years								
Radiation treatment	Observed	64	167	181	103	48	30	529
	Observed/Expected	0.73	0.69	0.87	0.79	0.74	0.71	0.77
No radiation treatment	Observed	27	165	129	39	15	12	360
	Observed/Expected	0.59	1.00	0.97	0.73	0.79	1.47	0.95

[a] Excluding >1 year

of the cervix also tend to have more children and to have had their first pregnancy at an earlier age than women in the general population (Kessler, 1976; Thomas, 1977): Thomas reported that 46% of cases (in-situ) compared with 30% of controls had first become pregnant before age 20. For breast cancer, a relative risk of about 2.5 might be expected for women who did not become pregnant before age 20 (MacMahon et al., 1973). The confounding effect can be estimated as

$$(0.46 + 0.54 \times 2.5)/(0.3 + 0.7 \times 2.5) = 0.88.$$

The risk for breast cancer might thus be expected to be about 12% lower among

Table 6. Multiple myeloma, stomach, colon and rectal cancer following irradiation for cancer of the cervix

Cancer		Time since irradiation (years)						
		<1	1–4	5–9	10–14	15–19	20+	Total [a]
Multiple myeloma	Observed	2	3	8	6	9	7	33
	Observed/Expected	0.63	0.31	0.84	0.83	2.08	2.14	0.97
Stomach	Observed	13	57	57	43	30	13	200
	Observed/Expected	0.55	0.87	0.96	1.00	1.20	0.71	0.95
Colon	Observed	15	81	88	70	42	33	314
	Observed/Expected	0.50	0.94	1.07	1.12	1.12	1.00	1.04
Rectum	Observed	9	36	43	55	31	32	197
	Observed/Expected	0.56	0.79	0.99	1.70	1.57	1.95	1.25

[a] Excluding <1 year

cervical cancer patients than average. The additional effect of parity on breast cancer risk after adjusting for age at first birth is not large (MacMahon et al., 1973; Tulinius et al., 1978) and would not be expected to confound the observations appreciably.

Similar arguments can be used to explain the excess of lung cancer among in-situ cases, bearing in mind that the prevalence of smoking is higher among cervical cancer patients than among the general population (see Day et al., 1983, p. 173).

A final point underlined by this study is that not only can cancer registries efficiently provide information on the issue in question, but they themselves should benefit from the study. Studies such as that described here provide a new way of testing many of the operational procedures of a registry. Identification of second cancers may have been defective; linkage to death records incomplete; treatment information, as recorded, may have been of even less value than originally imagined; or the criteria for deciding that a lesion is a second primary cancer may have been inadequate. In this last respect, an interesting finding in the study of cervical cancer was a gross excess of lung cancer registered as a second primary (Day et al., 1983, p. 167). The only convincing explanation of the size of this excess among women with invasive cervical cancer is that many metastases to the lung were misclassified. It is unlikely that attention would have been drawn to this procedural lapse unless a study of this type had been performed.

LEUKAEMIA ASSOCIATED WITH FIRST COURSE OF CANCER TREATMENT—US CANCER REGISTRIES

Many studies of the carcinogenicity of chemotherapeutic agents have concentrated on relatively small clinical series. This approach has, of course, been valuable, but it provides no perspective of the extent of iatrogenic cancer in the general population. The Surveillance, Epidemiology and End Results (SEER) programme in the USA records the first course of treatment of registered patients. In a recent study (Curtis et al., 1984) of leukaemia occurring as a second primary malignancy, risk was related

Table 7. Leukaemia following cancers of the breast, endometrium and ovary (SEER data, 1973–1980)

Site of initial cancer	Total number of patients		Type of treatment		
			Surgery alone	Radiotherapy no chemotherapy	Chemotherapy
Breast	59 115	Observed	26	10	6
		Expected	27.0	5.8	1.6
		Observed/Expected	0.96	1.72	3.75
Endometrium	20 846	Observed	2	14	0
		Expected	5.3	6.7	0.1
		Observed/Expected	0.38	2.09	–
Ovary	9 726	Observed	1	5	9
		Expected	1.2	0.5	1.0
		Observed/Expected	0.83	10.0	9.0
All initial	440 546	Observed	172	68	47
		Expected	174.6	49.3	20.8
		Observed/Expected	0.99	1.38	2.26

to the site of the prior malignancy and to the type of initial therapy. Hodgkin's disease as a prior malignancy was excluded from consideration by the SEER registration procedures. Although, naturally, some preleukaemias may have been missed and later courses of treatment not included, the study has clearly defined the magnitude of the problem, and the primary sites for which it arises. Some 440 000 cancer patients were included in the study, and 327 leukaemias identified as second primaries. Of these leukaemias, 47 occurred following chemotherapy for the initial primary, as opposed to 20.8 expected. For patients initially treated by radiotherapy, 68 leukaemias were subsequently recorded, as opposed to 40.3 expected. For those initially treated by surgery alone, 172 leukaemias were later registered, with 174.6 expected. These figures suggest that both follow-up procedures and the calculation of expected numbers were satisfactory. Almost the entire excess was confined to patients with initial primaries of the breast, endometrium and ovary (Table 7). It is to be hoped that this study will be followed by a case-control investigation to identify the specific chemotherapeutic agents involved. Although less detailed in its description of the risk associated with specific chemotherapeutic regimens than some studies based on large clinical trials, this investigation provides important and complementary information. Both the size of the problem and the initial primary sites affected are clearly delineated, and a foundation constructed for subsequent, more detailed studies.

REFERENCES

Boice, J.D. & Hutchison, G.B. (1980) Leukemia in women following radiotherapy for cervical cancer: ten-year follow-up of an international study. *J. natl Cancer Inst.*, **65**, 115–129

The Cancer Registry of Norway (1980) *Survival of Cancer Patients. Cases Diagnosed in Norway, 1968–1975,* Oslo

Curtis, R.E., Hankey, B.F., Myers, M.H. & Young, J.L. (1984) Risk of leukemia associated with the first course of cancer treatment: an analysis of the Surveillance, Epidemiology, and End Results Program experience. *J. natl Cancer Inst., 72,* 531–544

Day, N.E., Boice, J.D., Jr, Andersen, A., Brinton, L.A., Brown, R., Choi, N.W., Clarke, E.A., Coleman, M.P., Curtis, R.E., Flannery, J.T., Hakama, M., Hakulinen, T., Howe, G.R., Jensen, O.M., Kleinerman, R.A., Magnin, D., Magnus, K., Makela, K., Malker, B., Miller, A.B., Nelson, N., Patterson, C.C., Pettersson, F., Pompe-Kirn, V., Primic-Zakelj, M., Prior, P., Ravnihar, B., Skeet, R.G., Skjerven, J.E., Smith, P.G., Sok, M., Spengler, R.F., Storm, H.H., Tomkins, G.W.O. & Wall, C. (1983) Summary chapter. In: Day, N.E. & Boice, J.D., Jr, eds, *Second Cancer in Relation to Radiation Treatment for Cervical Cancer. A Cancer Registry Collaboration (IARC Scientific Publications No. 52),* Lyon, International Agency for Research on Cancer, pp. 137–181

Hakulinen, T., Pukkala, E., Hakama, M., Lehtonen, M., Saxen, E. & Teppo, L. (1981) Survival of cancer patients in Finland in 1953–1974. *Ann. clin. Res., 13,* Suppl. 31

Hutchison, G.B. (1968) Leukemia in patients with cancer of the cervix uteri treated with radiation. A report covering the first 5 years of an international study. *J natl Cancer Inst., 40,* 951–982

Kessler, I.I. (1976) Human cervical cancer as a venereal disease. *Cancer Res., 36,* 783–791

MacMahon, B., Cole, P. & Brown, J. (1973) Etiology of human breast cancer: a review. *J. natl Cancer Inst., 50,* 21–42

Malker, B. & Pettersson, F. (1983) *Second primary cancers after treatment for cervical cancer: a study of 58 731 Swedish women with invasive or in situ cancer.* In: Day, N.E. & Boice, J.D., Jr, eds, *Second Cancer in Relation to Radiation Treatment for Cervical Cancer. A Cancer Registry Collaboration (IARC Scientific Publications No. 52),* Lyon, International Agency for Research on Cancer, pp. 87–96

Stovall, M. (1983) *Organ doses from radiotherapy of cancer of the uterine cervix.* In: Day, N.E. & Boice, J.D., Jr, eds, *Second Cancer in Relation to Radiation Treatment for Cervical Cancer. A Cancer Registry Collaboration (IARC Scientific Publications No. 52),* Lyon, International Agency for Research on Cancer, pp. 131–136

Thomas, D.B. (1977) An epidemiologic study of carcinoma in situ and squamous dysplasia of the uterine cervix. *Am. J. Epidemiol., 98,* 10–28

Tulinius, H., Day, N.E., Johannesson, G., Bjarnason, O. & Gonzalez, M. (1978) Reproductive factors and risk for breast cancer in Iceland. *Int. J. Cancer, 21,* 724–730

11. THE ROLE OF CANCER REGISTRATION IN DEVELOPING COUNTRIES

C.L.M. OLWENY

University of Zimbabwe, School of Medicine, Avondale, Harare, Zimbabwe[1]

INTRODUCTION

To control cancer, it is vital that it be recognized as a problem and that the human and material resources be available to deal with it. Most developing countries have not as yet formulated national cancer policies, and cancer is not recognized as a major problem needing attention. According to the World Health Organization (1977), for individuals who have survived the first five years of life, cancer is one of the three major causes of death in both the more developed world and less developed countries. Numerically, the majority of the world's cancer patients are in the developing countries. This statement is based on estimated cancer incidence rates of 260 and 102 per 100 000 population for developed and developing countries, respectively (Parkin *et al.*, 1984), given the fact that 75% of the world's 4220 million people are to be found in developing countries (World Health Organization, 1977). On the basis of these figures, it is estimated that 2.84 million cancer cases occur each year in developed countries and 3.03 million cases per year in the less developed countries. Since, in addition, the age structure of the population in developing countries is changing rapidly, largely because of a reduction in mortality from infectious disease, together with the adoption of 'western' life-styles, the cancer risk in developing countries is certain to increase.

In spite of these alarming facts, most countries in the developing world lack appropriate cancer control delivery systems.

To tackle any problem, it is essential first to assess its nature and size. The establishment of cancer registries in developing countries is of prime importance if any meaningful control programmes are to be launched.

---

[1] Present address: Medical Oncology, Royal Adelaide Hospital, North Terrace, Adelaide, SA 5000, Australia.

ROLE OF CANCER REGISTRATION IN GENERAL

A cancer registry can be defined as a facility for the collection, storage, analysis and interpretation of data on persons with cancer (Muir & Nectoux, 1977). The value of cancer registration in any cancer control programme cannot be over-emphasized. If properly undertaken, cancer registration can:

(1) facilitate the assessment of cancer incidence in its operational area, and, by defining the extent of malignant disease, provide information needed for planning control measures;

(2) provide a data base for epidemiological research which could suggest possible etiological factors; and

(3) provide information on treatment efficacy and the value of any preventive measures undertaken.

Thus, cancer registration gives guidance on past, present and future trends as well as on needs for cancer control measures.

CANCER PATTERNS AND TRENDS

Of an estimated 50 million deaths a year in the world, about 4.3 million are attributed to cancer, and 2.3 million of these occur in developing regions. In Europe and North America, about 20% of the population will die of cancer if the present mortality trends are maintained. WHO has estimated that by the year 2000 the number of cancer deaths throughout the world may rise by 50%, to approximately 8 million annually (World Health Organization, 1977). These estimates are based on observable changes in the health spectrum and demographic structure of the world population. Successful efforts have been made in developing countries to reduce the numbers of premature deaths, and the life expectancy can be expected to continue to rise; the decline in infant mortality coupled with decreased fertility in many developing countries will result in a rise in population in the high-risk age groups for cancer both in absolute and relative terms. With socio-economic development, life style and behaviour patterns are changing, and these also may be associated with an increased risk of certain cancers, and possibly decreased risks of others. Important changes can be expected in the environment consequent upon urbanization and industrialization, some of which will cause an increase in cancer risk for certain populations. Unfortunately, reliable data on cancer incidence, gathered by standard methods, are published for only about 5% of the world's population. Information on a further 10% is potentially available but has not, for various reasons, been included in appropriate publications (Waterhouse *et al.,* 1982). For instance, that latest issue of *Cancer Incidence in Five Continents* has very little information on Africa: the only information included is on Senegal. Data from some of the oldest registries in tropical Africa, like those of Kyadondo, Ibadan and Bulawayo, which have contributed immensely to our present knowledge of cancer patterns in Africa, were not included because of a general lack of current denominators from which estimates of incidence rates could be calculated. There are also major differences in the coverage of various continents and geographic regions: in 1975, figures on cancer mortality available to the WHO covered 78% and nearly 100%, respectively, of the populations of the Americas and Europe; while for Africa and Asia, data were available for less than 10% of the population—well below the world average figure of 36%. In Africa, mortality data are available only for Mauritius, Egypt, Cape Verde and South Africa, which account for only 15% of the total population. This is because registration of births, marriages and deaths is not mandatory in most African societies and nations.

In addition, the quality of diagnostic data reported on death certificates in developing countries is questionable. Only a small fraction of deaths are notified, and death certificates are generally completed by non-medical personnel; even when they are completed by medical personnel, the medical history may not be well known by the certifying physician. Furthermore, numerous errors are introduced during coding of death certificates. Thus, nearly half of deaths are ascribed to senility and ill-defined conditions, giving an artificially low figure for cancer mortality.

CANCER IS COMMON IN DEVELOPING COUNTRIES

Health workers in developing countries continue to be inundated by patients with infectious, parasitic and nutritional diseases. Consequently, a myth has been perpetuated that cancer is a disease of industrialization and therefore rare in developing countries. Even if this were true, recent data from China and Singapore should provide a warning of possible changes in disease patterns over a relatively short period of time. Within one generation, an observable change occurred from a pattern predominantly of infectious, parasitic and nutritional diseases to one of a preponderance of non-communicable diseases, such that cancer is now a major public health concern in those two countries. The same change can be expected to occur elsewhere with similar speed.

Such changes would be difficult to observe were it not for the existence of cancer registries. Data from existing population-based cancer registries indicate that age-specific incidence rates for cancer in developing countries are not much lower than those in the developed world. Frequently, however, there are marked differences in the kinds of cancer observed: cancer of the cervix and primary liver cancer are very common in developing countries but are relatively rare in Europe and North America. The data so far available suggest that in developing countries cancers afflict mainly young individuals, although this may merely reflect the generally young age of the population at risk. There have also been reports of observable trends with time: epitheliomas complicating tropical ulcers are on the decline in certain parts of Africa (Samitz, 1980), perhaps due to improvements in socio-economic status with the increasing availability of soap and clean water. Whatever the explanation for the differences in types of cancer observed—the young age at presentation or the changing trends—all stress the need for further, accurate documentation of observed features.

ROLE OF CANCER REGISTRATION IN DEVELOPING COUNTRIES

In developing countries, cancer registration is of great value in providing information to health planners. This information enables the planners to determine relative needs and the location and quality of services to control cancer vis-à-vis other diseases (Muir & Nectoux, 1977). On the one hand, in a country where many chemosensitive tumours (e.g., acute leukaemia, choriocarcinoma and Burkitt's lymphoma) are encountered, emphasis might be placed on chemotherapeutic facilities rather than on radiotherapy. On the other hand, if the tumours commonly dealt with are radiosensitive, like nasopharyngeal and breast carcinomas, then radiotherapy would be more useful. When an environmental cause can be identified for common malignancies (e.g., hepatocellular carcinoma, lung cancer), then greater emphasis might be focussed on primary prevention.

Cancer registries endeavour to record all newly diagnosed cases of cancer, stratifying them by age, sex, anatomical site, area of residence, occupation, religion and ethnic group. Analysis of this information may reveal possible variations in tumour distribution or clusters within a given country and thus permit the formulation of etiological hypotheses. The high prevalence of endemic Kaposi's sarcoma in predominantly Negroid South Sudan in comparison with the Arabic North may point to a possible genetic influence, although environmental factors may equally explain the difference. The high frequency of nasopharyngeal and oesophageal carcinomas in people living around Lake Victoria might suggest environmental factors associated with the lake; nasopharyngeal carcinoma has been linked with the consumption of salted dried fish in China.

Cancer registration provides good baseline data on the situation before an intervention programme is initiated; if monitoring is continued during and after the intervention period, progress achieved can be more readily assessed. The planned joint IARC/MRC and Government of Gambia hepatitis intervention study (IARC, 1984), which aims to compare the cancer outcome of 30 000 immunized male infants with 30 000 controls will take over 30 years to demonstrate an 80% efficacy of the vaccine in preventing liver cancer. A cancer registry will be set up to monitor the occurrence of all new liver cancer cases over this period. Similarly, in Swaziland, the UNEP/IARC programme (IARC, 1984) aims to correlate the agricultural improvements made to control aflatoxin contamination of food in different parts of the country with liver cancer incidence in the same areas. The Swaziland Cancer Registry has been used to monitor liver cancer frequency by geographic area, and it is hoped that this activity will continue.

Good registries may be used to discover possible time trends. For instance, a declining incidence of Burkitt's lymphoma has been observed in endemic areas of East and West Africa. Studies in the North Mara region of Tanzania (Geser & Brubaker, 1985) and in the West Nile district of Uganda (Williams, 1985) clearly indicate a definite time trend for Burkitt's lymphoma (Figure 1). The explanation for this decline in incidence over the last decade or so is not immediately clear, but may perhaps be ascribed to elimination of an environmental factor or factors in the regions studied. As stated earlier, the decline in the numbers of epitheliomas in Tanzania has been

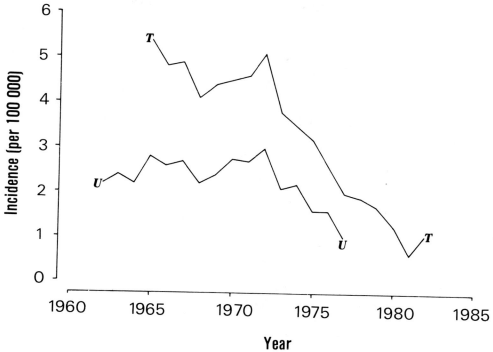

Fig. 1. Time trends in incidence of Burkitt's lymphoma in East Africa; three-year moving average T, North Mara (Tanzania), from Geser & Brubaker (1985); U, West Nile (Uganda), from Williams (1985)

ascribed to the decline in the incidence of tropical ulcers owing to greater availability of soap, water and antibiotics (Samitz, 1980).

Cancer registration can provide useful evaluation of treatment facilities offered. Although survival rate and remission durations for various types of cancers can be determined by means of controlled clinical trials, cancer registry evaluation of case management for all persons with cancer, whether treated or not, gives a more representative picture of survival of all cancer victims (Waterhouse, 1974). Cancer registration calls for good record keeping; this in itself results in improvement in patient care. In addition, the centralization of records of cancer patients facilitates retrieval of patient information and hence improved patient management. Follow-up, which is included in the activities of some cancer registries, provides an excellent opportunity to assess treatment results and to determine the natural history of malignant disease.

The natural history of malignancies in developing countries may well be different from that in other parts of the world. Cancer of the liver in developing countries runs a very rapid downhill course: most patients who are not treated are dead within three months of diagnosis. This is at variance with the situation observed in Europe or North America, where primary liver cancer runs a relatively slow course, and 5–10%

of victims are still alive at 12 months. Although the difference may depend simply on delays in reporting to hospitals in developing countries, it is also possible that when these tumours occur in individuals who are already 'run down' by nutritional, parasitic and infectious diseases they behave differently; in such situations, tumours may behave opportunistically, taking advantage of the immune-compromised host.

Finally, it is conceivable that morphologically similar tumours have different pathogeneses. The epidemiology and clinical course of endemic Kaposi's sarcoma are clearly different from those of epidemic forms associated with acquired immunodeficiency syndrome, and yet, morphologically, the two diseases are very similar. It is only through accurate and continued registration of these differences that explanations can emerge.

PROBLEMS OF CANCER REGISTRATION IN DEVELOPING COUNTRIES

Cancer registration is not without problems (Grundmann & Pedersen, 1975), and in developing countries, the problems are compounded by the recent world recession and harsh climatic conditions (prolonged droughts and floods) which have necessitated reallocation of resources. Although it is hazardous to generalize, because of great differences in the nature and quality of health information systems, most developing countries share certain common problems related to cancer statistics. The major problems include:

(1) *Lack of basic health services*

In order for a cancer registry to be valid, reporting must be as complete as possible. This is difficult to achieve in areas where medical services are scarce and unevenly distributed. The vast majority of patients in developing countries are still treated by traditional healers, who have little or no knowledge of cancer. The few health personnel and services available are concentrated in cities and urban centres, although, paradoxically, the majority of the population is still to be found in rural areas. The few clinics that exist are perpetually overcrowded, and busy health care workers may not appreciate the justifications for diverting the scarce resources for a detailed examination and investigation. The additional burden of notifying cases may discourage regular reporting of those identified.

The quality of any registry is dependent on the criteria upon which diagnosis is accepted. The frequency of superficial cancers, like those of the skin, penis or cervix, may be over-represented because of the easy accessibility of these organs to biopsy. Internal malignancies like those of liver and pancreas are likely to be under-reported (Burkitt *et al.*, 1968). Potentially hazardous procedures such as liver biopsy tend to be avoided, and since noninvasive techniques (e.g., ultrasound, isotopic scans and computerized axial tomography) are beyond the economic reach of most health budgets, diagnoses of liver, pancreatic and brain tumours are rarely confirmed. Postmortem examinations may not be acceptable for social, cultural or religious reasons, and many diagnoses may remain speculative.

(2) *Lack of demographic data*

There are no reliable figures for the total population in most communities in the developing world. When they exist, they are sometimes not stratified for appropriate variables such as age, sex and ethnic group. Besides, the populations tend to be unstable: a number of communities in some developing countries are still nomadic; other communities may be forced to move because of social, political or economic upheavals; rural people tend to migrate to cities in search of employment and higher living standards, some of whom may become permanent dwellers in periurban slums with no fixed addresses; some return to rural areas, either having earned enough or because they have become disenchanted with city life. Unfortunately, there are no records of such movements.

Apart from the problem of unstable populations, it is often difficult to identify individuals unequivocally: few countries in the developing world have instituted systems of national registration; registration of births and deaths is not compulsory; social security systems, through which individuals may be identified, are non-existent; some people change names at will; those giving birth to, or who have fathered, twins acquire new and prestigious titles and names that do not correspond to any other, so that the unwary registrar may record such individuals more than once. There is thus no reliable population denominator.

(3) *Lack of appropriately trained personnel*

Perhaps the major problem in the development and maintenance of cancer statistics in developing countries is the lack of appropriately and adequately trained persons to fulfill the necessary functions in registries. The collecting, storage, analysis, interpretation and utilization of cancer data require a high level of training at a variety of stages. In some developing countries, there may not be a resident pathologist to help with histological diagnoses. In others there may be no epidemiologists and/or statisticians capable of analysing and interpreting the data. Trained oncologists are even rarer: it has been estimated that, in Africa, and this is probably true for most developing countries, there is at present one qualified oncologist for every 10 million population. It is obvious, therefore, that the data collected may not be usefully applied. Often, data cannot be analysed for lack of equipment; and if it is, it may not reach the health planners.

(4) *Lack of adequate follow-up*

Follow-up is desirable for good clinical management, and good clinical information gives credence to cancer statistics. However, follow-up tends to be difficult and at best fragmented in developing countries because of poor communications and frequent social and political upheavals. Roads are impassable for most of the year; telephones, when they exist, are more often than not out of order; postal systems are unreliable, to say the least, so much so that the tracing of a patient becomes a nightmare. Further, most patients do not appreciate the need for follow-ups; others would be willing to report back as requested but for the expense involved: many have to sell their only means of livelihood to obtain money to return on the appointed date.

Since few have permanent addresses, it becomes impossible to seek information by mail.

PROSPECTS FOR THE FUTURE

Recognizing that cancer is numerically as common, if not more common, in developing countries as in the developed nations, it is imperative that efforts be made to develop or to plan urgently cancer control programmes in developing countries. Because of the variations in facilities and both material and human resources in different countries, it would be unwise to make generalizations or to extrapolate experiences gained in one country to another. However, certain strategies probably apply equally to most developing countries. If cancer control is to be attained, it is imperative that each country pursue as a matter of urgency some guidelines, among which are:

(1) *Formulation of a national cancer policy*

Many countries in the Third World have no cancer policy, and cancer as such is not recognized as a problem of priority. WHO has over the years assisted some member states to identify their needs in the cancer field, to overcome inadequate management practices and to support the relevant part of the health sector.

Experience gained in certain developing countries indicates that the management aspects of national policy formulation are as important as the technical aspects. Lack of central management, multidisciplinary approaches and manpower and delays in mobilizing personnel to assume leadership have largely accounted for any lack of progress. Observations in Sri Lanka indicate the feasibility of utilizing primary health workers to control oral cancer and suggest that this approach may be cost effective (Warnakulasuriya *et al.*, 1984). Similarly, traditional birth attendants may be used for delivering vaccinations, especially hepatitis B vaccine which must be given soon after birth.

(2) *Attainment of reliable and stable demographic data*

Every effort should be made to secure reliable and stable demographic data. Otherwise, it is impossible to develop cancer incidence rates, since these are derived from population-based registries which demand accurate baseline population data (MacLennan *et al.*, 1978).

(3) *Reactivation of old registries and establishment of new ones*

A number of long-established cancer registries, like those in Bulawayo, Ibadan and Kyadondo, have contributed immensely to the understanding of cancer patterns in Africa and elsewhere. Efforts ought to be made to reanimate these registries and to ensure that the quality of their data is maintained. However, there are few registers in developing countries, and attempts should be made to establish at least one in each country. It may not be feasible, given current constraints, to establish population-based registries; however, good hospital- or pathology-based registries might be useful until such time as the situation permits country-wide population registries.

The value of cancer registries cannot be over-emphasized, given the situation in most developing countries where national death statistics are either not available or tend to be of poor quality. The establishment of registries is relatively easy and cheap, and if they are located in a major hospital or university with diagnostic facilities, the quality of the information generated can be good and reliable, even though hospital-based registries tend to cater for a select few—usually urban and periurban dwellers—and for an undefined population of temporary migrants from rural areas.

(4) Training of personnel

Lack of appropriately trained personnel has hampered development in many fields in the Third World. In oncology, there is a dire shortage of oncologists, epidemiologists, pathologists, cancer registrars and statisticians capable of collecting, storing, analysing, interpreting and utilizing cancer data.

Training at all levels, including traditional healers and primary health care workers, is a prerequisite if the final goal of cancer control is to be attained by the year 2000.

REFERENCES

Burkitt, D.P., Hutt, M.S.R. & Slavin, G. (1968) Clinicopathological studies of cancer distribution in Africa. *Br. J. Cancer, 22,* 1–6

Geser, A. & Brubaker, G. (1985) *A preliminary report of epidemiological studies of Burkitt's lymphoma, Epstein-Barr virus infection and malaria in North Mara, Tanzania.* In: Lenoir, G., O'Conor, G. & Olweny, C.L.M., eds, *Burkitt's Lymphoma: A Human Cancer Model (IARC Scientific Publications No. 60),* Lyon, International Agency for Research on Cancer (in press)

Grundmann, E. & Pedersen, E., eds (1975) Cancer registry. *Recent Results Cancer Res., 50*

International Agency for Research on Cancer (1984) *Annual Report 1984,* Lyon

MacLennan, R., Muir, C.S., Steinitz, R. & Winkler, A. (1978) *Cancer Registration and its Techniques (IARC Scientific Publications No. 21),* Lyon, International Agency for Research on Cancer

Muir, C.S. & Nectoux, J. (1977) Role of the cancer registry. *Natl Cancer Inst. Monogr., 47,* 3–6

Parkin, D.M., Stjernswärd, J. & Muir, C.S. (1984) Estimates of the worldwide frequency of twelve major cancers. *Bull. World Health Organ., 62,* 163–182

Samitz, M.H. (1980) Dermatology in Tanzania; problems and solutions. *Int. J. Dermatol., 19,* 102–106

Warnakulasuriya, K.A.A.S., Ekanayake, A.N.I., Sivayoham, S., Stjernswärd, J., Pindborg, J.J., Sobin, L.H. & Perera, K.S.G.P. (1984) Utilization of primary health care workers for early detection of oral cancer and precancer cases in Sri Lanka. *Bull. World Health Organ., 62,* 243–250

Waterhouse, J.A. (1974) *Cancer Handbook of Epidemiology and Prognosis*, Edinburgh, Churchill Livingstone

Waterhouse, J., Muir, C.S., Shanmugaratnam, K. & Powell, J., eds (1982) *Cancer Incidence in Five Continents Vol. IV (IARC Scientific Publications No. 42)*, Lyon, International Agency for Research on Cancer

Williams, E.H. (1985) In: *Cancer Occurrence in Developing Countries (IARC Scientific Publication)*, Lyon, International Agency for Research on Cancer (in preparation)

World Health Organization (1977) *World Health Statistics Annual 1977*, Geneva

SUBJECT INDEX

Birth cohort analysis, 34–35
Bladder, cancer of, 33, 67, 71, 93, 94, 100, 117
Bone tumour, registry of, 9
  cancer of, 20
Brain, cancer of, 28, 30, 148
Breast, cancer of, 15, 16, 17, 19, 28, 31, 36, 41, 92, 93, 94, 95, 96, 97, 98, 101, 102, 116, 117, 118, 127, 128, 136, 138–140, 146
  screening of, 53–54, 58, 59, 83
Burkitt's lymphoma, 146, 147

Cancer registry
  for epidemiology, 7, 9, 16–18, 144
  history of, 3–12
  hospital-based, 13, 118, 121–131, 150–151
  population-based, 13, 17, 27, 29, 53, 55, 56, 58, 59, 79, 87, 88, 89, 92–105, 109–110, 112, 122, 125, 129, 145, 150
  role, 143–144
Case-control study, 17, 21, 22, 37, 52, 58, 65, 66, 70–71, 135, 140
Census, 3, 5, 6, 15, 16
Cervix uteri, cancer of, 7, 15, 16, 20, 28, 41–42, 93, 94, 95, 96, 97, 98, 101, 102, 134–139, 145, 148
  screening, 46–53, 56, 57–59, 83
Chemotherapy, 15, 116, 128
  carcinogenicity of, 139–140, 146
Choriocarcinoma, 14, 82, 146
Classification of tumours, 67, 87, 88, 114
Clinical trial, 14, 109–110, 114, 118, 121, 128, 133, 140, 146
Clusters of cancer cases, 36, 67, 146
Coding, 19, 20, 87, 113, 124, 129, 145
Cohort study, 17, 18, 21, 36–37, 50, 65, 66, 69–70, 71
Colon, cancer of, 15, 17, 19, 20, 28, 32, 33, 36, 94, 95, 96, 97, 98, 100, 101, 102, 128, 136, 139
Compulsory registration, 6, 8, 9
Computer, 22, 37, 115, 129
  simulation, 37–38, 55
Confidentiality, 21–23, 24, 71–72, 129
Corpus uterus, cancer of, 7, 16, 32, 33, 94

Data protection, 65, 72
Death certificate, 7, 8, 14, 19, 21, 22, 23, 68, 78, 89, 145
  accuracy of cause of death, 20, 27, 90–91
Denominators, 16, 67–69, 144
Developing countries, 143–152
Diagnosis, 14, 16, 17, 20, 84, 87, 92, 114, 122, 124, 125, 127, 128, 129, 145, 148
Diet, 17, 18, 23, 36, 66

Endometrium, cancer of, 18, 20, 140
Epithelioma, 145, 146

Follow-up (notice), 8, 14, 16, 36–37, 69, 88–89, 114, 115, 122, 124, 126–127, 129, 135, 140, 147, 149
Funding, 23, 71, 76, 122

Gall-bladder, cancer of, 33
Gastrointestinal tumour, registry of, 9
Glioma, 117
Gynaecological tumour, registry of, 9

Health education, 42, 84, 127
Hepatocellular carcinoma, 17, 146
Histological type, 20, 41, 87, 92, 100–101, 103, 124, 127
Hodgkin's disease, 15, 19, 84, 103
Hospital discharge record, 19

Identification of individuals, viii, 124, 149
Incidence rates, vii, 8, 17, 18, 19, 27, 32, 36, 39–41, 47, 48, 50, 55, 58, 75, 84, 144, 150
Industrial risk, 65–73
International Association of Cancer Registries, 9–10

Job-exposure matrix, 69

Kaposi's sarcoma, 146, 148
Kidney, cancer of, 84

Large bowel, cancer of, 16, 54
Larynx, cancer of, 20, 36, 84, 93, 94, 95, 100
'Lead time', 46, 50, 54, 102
'Length bias', 46, 54
Leukaemia, 20, 28, 30, 36, 82, 83, 134, 135, 138, 139–140
  acute lymphatic, 15, 84, 103, 117, 135, 138, 146
  registry of, 9
Lifestyle, 17, 23, 144
Life table, for computing survival, 89, 90, 91
Lip, cancer of, 19, 36
Liver, cancer of, 28, 93, 94, 99, 100, 145, 146, 147, 148
Lung, cancer of, 8, 15, 17, 19, 20, 23, 28, 36, 37–41, 46, 54, 67, 69, 70, 71, 90, 93, 94, 95, 96, 97, 98, 100, 101, 102, 104, 117, 139, 146
Lymphoma, 28, 30, 82

Maps, cancer incidence, 30–31
Melanoma, 33, 34, 35, 117
Morbidity statistics, 6, 7, 8
Mortality rates, vii, 4, 5, 8, 17, 19, 41, 48, 50, 53, 54, 59, 68, 69, 78, 83, 144–145
Multiple tumours, 36, 69, 89
Myeloma, 117, 136, 139

Nasal cavity, cancer of, 65, 67–68, 70
Nasopharynx, cancer of, 146

Occupation, vii, 7, 8, 15, 16, 21, 31, 37, 65–73, 125–126
Oesophagus, cancer of, 93, 94, 99, 100, 146
Oral cavity, cancer of, 45, 59, 117, 150
Ovary, cancer of, 93, 94, 117, 118, 140

Paediatric tumours, registry of, 9
Pancreas, cancer of, 20, 93, 94, 100, 148
Penis, cancer of, 148
Personal number, 21, 37, 71, 117
Prevalence, 28, 29

SUBJECT INDEX

Prevention of cancer, vii, 27–44
    evaluation of, 37–42
Priorities, identification of, 27–35, 76
Proportional analysis, 68
Prostate, cancer of, 19, 28, 29, 30, 93, 94, 117

Radiotherapy, 15, 16, 37, 80–82, 110, 116, 134–140, 146
Rectum, cancer of, 20, 94, 99, 100, 117, 128, 136, 139
Rehabilitation, 14, 125
Relative frequency, vii
Retinoblastoma, 82
    registry of, 9
Risk factor, 35–37, 65–73

Screening programme, viii, 23, 45–63, 83, 87, 103, 125, 126
    evaluation of, 16, 41–42, 45–63
    planning, 55–57, 110
Second primary tumour, 20, 133–141
SEER, 15, 23, 122–123, 128, 139–140
Services for cancer patient, 75–84
Skin, cancer of, 28, 148
Smoking, 37–41, 66, 139
Social status, 7, 15, 31, 32, 36, 52, 68, 92, 101, 144, 145
Staffing, 23, 71, 129, 149
Stage distribution, 55, 58, 59, 84, 87, 95, 99, 114, 124–125
Stomach, cancer of, 7, 15, 16, 28, 32, 33, 34, 35, 54–55, 93, 94, 95, 96, 97, 98, 100, 101, 102, 103, 104, 105, 117, 136, 139
Survival, viii, 14–15, 17, 19, 42, 55, 58, 59, 77, 79, 87–107, 114, 118, 127–128, 146

Testis, cancer of, 15, 16, 19, 20, 28
Thyroid, cancer of, 28
Time trends, 16, 32–35, 48, 58, 87, 102–103, 104, 144–145, 146, 147
Training, 123, 127, 151
Treatment, 14, 15, 37, 83, 92, 103, 109–119, 127, 144, 146
    carcinogenic effects of, 36, 133–141

Urban-rural difference, 31, 32

Vital statistics, office of, 18, 19, 20, 21
Voluntary reporting, 7, 8, 9

IARC MONOGRAPHS ON THE EVALUATION OF THE CARCINOGENIC RISK OF CHEMICALS TO HUMANS
(English editions only)

(Available from WHO Sales Agents)

Volume 1
Some inorganic substances, chlorinated hydrocarbons, aromatic amines, N-nitroso compounds, and natural products (1972)
184 pp.; out of print

Volume 2
Some inorganic and organometallic compounds (1973)
181 pp.; out of print

Volume 3
Certain polycyclic aromatic hydrocarbons and heterocyclic compounds (1973)
271 pp.; out of print

Volume 4
Some aromatic amines, hydrazine and related substances, N-nitroso compounds and miscellaneous alkylating agents (1974)
286 pp.; US$7.20; Sw.fr. 18.-

Volume 5
Some organochlorine pesticides (1974)
241 pp.; out of print

Volume 6
Sex hormones (1974)
243 pp.; US$7.20; Sw.fr. 18.-

Volume 7
Some anti-thyroid and related substances, nitrofurans and industrial chemicals (1974)
326 pp.; US$12.80; Sw.fr. 32.-

Volume 8
Some aromatic azo compounds (1975)
357 pp.; US$14.40; Sw.fr. 36.-

Volume 9
Some aziridines, N-, S- and O-mustards and selenium (1975)
268 pp.; US$10.80; Sw.fr. 27.-

Volume 10
Some naturally occurring substances (1976)
353 pp.; US$15.00; Sw.fr. 38.-

Volume 11
Cadmium, nickel, some epoxides, miscellaneous industrial chemicals and general considerations on volatile anaesthetics (1976)
306 pp.; US$14.00; Sw.fr. 34.-

Volume 12
Some carbamates, thiocarbamates and carbazides (1976)
282 pp.; US$14.00; Sw.fr. 34.-

Volume 13
Some miscellaneous pharmaceutical substances (1977)
255 pp.; US$12.00; Sw.fr. 30.-

Volume 14
Asbestos (1977)
106 pp.; US$6.00; Sw.fr. 14.-

Volume 15
Some fumigants, the herbicides 2,4-D chlorinated dibenzodioxins and miscellaneous industrial chemicals (1977)
354 pp.; US$20.00; Sw.fr. 50.-

Volume 16
Some aromatic amines and related nitro compounds - hair dyes, colouring agents and miscellaneous industrial chemicals (1978)
400 pp.; US$20.00; Sw.fr. 50.-

Volume 17
Some N-nitroso compounds (1978)
365 pp.; US$25.00; Sw.fr. 50.-

Volume 18
Polychlorinated biphenyls and polybrominated biphenyls (1978)
140 pp.; US$13.00; Sw.fr. 20.-

IARC MONOGRAPHS SERIES

Volume 19
Some monomers, plastics and synthetic elastomers, and acrolein (1979)
513 pp.; US$35.00; Sw.fr. 60.-

Volume 20
Some halogenated hydrocarbons (1979)
609 pp.; US$35.00; Sw.fr. 60.-

Volume 21
Sex hormones (II) (1979)
583 pp.; US$35.00; Sw.fr. 60.-

Volume 22
Some non-nutritive sweetening agents (1980)
208 pp.; US$15.00; Sw.fr. 25.-

Volume 23
Some metals and metallic compounds (1980)
438 pp.; US$30.00; Sw.fr. 50.-

Volume 24
Some pharmaceutical drugs (1980)
337 pp.; US$25.00; Sw.fr. 40.-

Volume 25
Wood, leather and some associated industries (1981)
412 pp.; US$30.00; Sw.fr. 60.-

Volume 26
Some antineoplastic and immuno-suppressive agents (1981)
411 pp.; US$30.00; Sw.fr. 62.-

Volume 27
Some aromatic amines, anthraquinones and nitroso compounds, and inorganic fluorides used in drinking-water and dental preparations (1982)
341 pp.; US$25.00; Sw.fr. 40.-

Volume 28
The rubber industry (1982)
486 pp.; US$35.00; Sw.fr. 70.-

Volume 29
Some industrial chemicals and dyestuffs (1982)
416 pp.; US$30.00; Sw.fr. 60.-

Volume 30
Miscellaneous pesticides (1983)
424 pp; US$30.00; Sw.fr. 60.-

Volume 31
Some food additives, feed additives and naturally occurring substances (1983)
314 pp.; US$30.00; Sw.fr. 60.-

Volume 32
Polynuclear aromatic compounds, Part 1, Environmental and experimental data (1984)
477 pp.; US$30.00; Sw.fr. 60.-

Volume 33
Polynuclear aromatic compounds, Part 2, Carbon blacks, mineral oils and some nitroarene compounds (1984)
245 pp.; US$25.00; Sw.fr. 50.-

Volume 34
Polynuclear aromatic compounds, Part 3, Some complex industrial exposures in aluminium production, coal gasification, coke production, and iron and steel founding (1984)
219 pages; US$20.00; Sw.fr. 48.-

Volume 35
Polynuclear aromatic compounds, Part 4, Bitumens, coal-tars and derived products, shale-oils and soots (1985)
271 pages; US$25.00; Sw.fr. 70.-

Supplement No. 1
Chemicals and industrial processes associated with cancer in humans (IARC Monographs, Volumes 1 to 20) (1979)
71 pp.; out of print

Supplement No. 2
Long-term and short-term screening assays for carcinogens: a critical appraisal (1980)
426 pp.; US$30.00; Sw.fr. 60.-

Supplement No. 3
Cross index of synonyms and trade names in Volumes 1 to 26 (1982)
199 pp.; US$30.00; Sw.fr. 60.-

Supplement No. 4
Chemicals, industrial processes and industries associated with cancer in humans (IARC Monographs, Volumes 1 to 29) (1982)
292 pp.; US$30.00; Sw.fr. 60.-

INFORMATION BULLETIN ON THE SURVEY OF CHEMICALS BEING TESTED FOR CARCINOGENICITY
No. 8 (1979)
Edited by M.-J. Ghess, H. Bartsch & L. Tomatis
604 pp.; US$20.00; Sw.fr. 40.-

INFORMATION BULLETIN ON THE SURVEY OF CHEMICALS BEING TESTED FOR CARCINOGENICITY
No. 9 (1981)
Edited by M.-J. Ghess, J.D. Wilbourn, H. Bartsch & L. Tomatis
294 pp.; US$20.00; Sw.fr. 41.-

INFORMATION BULLETIN ON THE SURVEY OF CHEMICALS BEING TESTED FOR CARCINOGENICITY
No. 10 (1982)
Edited by M.-J. Ghess, J.D. Wilbourn, H. Bartsch
326 pp.; US$20.00; Sw.fr. 42.-

INFORMATION BULLETIN ON THE SURVEY OF CHEMICALS BEING TESTED FOR CARCINOGENICITY
No. 11 (1984)
Edited by M.-J. Ghess, J.D. Wilbourn, H. Vainio & Bartsch
336 pp.; US$20.00; Sw.fr. 48.-

PUBLICATIONS OF THE INTERNATIONAL AGENCY FOR RESEARCH ON CANCER

SCIENTIFIC PUBLICATIONS SERIES

(Available from Oxford University Press)

No. 1 LIVER CANCER (1971)
176 pages; £10-

No. 2 ONCOGENESIS AND HERPES VIRUSES (1972)
Edited by P.M. Biggs, G. de Thé & L.N. Payne, 515 pages; £30.-

No. 3 N-NITROSO COMPOUNDS - ANALYSIS AND FORMATION (1972)
Edited by P. Bogovski, R. Preussmann & E.A. Walker, 140 pages; £8.50

No. 4 TRANSPLACENTAL CARCINOGENESIS (1973)
Edited by L. Tomatis & U. Mohr, 181 pages; £11.95

No. 5 PATHOLOGY OF TUMOURS IN LABORATORY ANIMALS. VOLUME 1. TUMOURS OF THE RAT. PART 1 (1973)
Editor-in-Chief V.S. Turusov,
214 pages; £17.50

No. 6 PATHOLOGY OF TUMOURS IN LABORATORY ANIMALS. VOLUME 1. TUMOURS OF THE RAT. PART 2 (1976)
Editor-in-Chief V.S. Turusov
319 pages; £17.50

No. 7 HOST ENVIRONMENT INTERACTIONS IN THE ETIOLOGY OF CANCER IN MAN (1973)
Edited by R. Doll & I. Vodopija,
464 pages; £30.-

No. 8 BIOLOGICAL EFFECTS OF ASBESTOS (1973)
Edited by P. Bogovski, J.C. Gilson, V. Timbrell & J.C. Wagner,
346 pages; £25.-

No. 9 N-NITROSO COMPOUNDS IN THE ENVIRONMENT (1974)
Edited by P. Bogovski & E.A. Walker
243 pages; £15.-

No. 10 CHEMICAL CARCINOGENESIS ESSAYS (1974)
Edited by R. Montesano & L. Tomatis,
230 pages; £15.-

No. 11 ONCOGENESIS AND HERPESVIRUSES II (1975)
Edited by G. de-Thé, M.A. Epstein & H. zur Hausen
Part 1, 511 pages; £30.-
Part 2, 403 pages; £30.-

No. 12 SCREENING TESTS IN CHEMICAL CARCINOGENESIS (1976)
Edited by R. Montesano, H. Bartsch & L. Tomatis, 666 pages; £30.-

No. 13 ENVIRONMENTAL POLLUTION AND CARCINOGENIC RISKS (1976)
Edited by C. Rosenfeld & W. Davis
454 pages; £17.50

No. 14 ENVIRONMENTAL N-NITROSO COMPOUNDS - ANALYSIS AND FORMATION (1976)
Edited by E.A. Walker,
P. Bogovski & L. Griciute, 512 pages; £35.-

No. 15 CANCER INCIDENCE IN FIVE CONTINENTS. VOL. III (1976)
Edited by J. Waterhouse, C.S. Muir, P. Correa & J. Powell, 584 pages; £35.-

No. 16 AIR POLLUTION AND CANCER IN MAN (1977)
Edited by U. Mohr, D. Schmahl & L. Tomatis, 331 pages; £30.-

No. 17 DIRECTORY OF ON-GOING RESEARCH IN CANCER EPIDEMIOLOGY 1977 (1977)
Edited by C.S. Muir & G. Wagner,
599 pages; out of print

SCIENTIFIC PUBLICATIONS SERIES

No. 18 ENVIRONMENTAL CARCINO-
GENS - SELECTED METHODS OF
ANALYSIS
Editor-in-Chief H. Egan
Vol. 1 - ANALYSIS OF VOLATILE
NITROSAMINES IN FOOD (1978)
Edited by R. Preussmann,
M. Castegnaro, E.A. Walker
& A.E. Wassermann, 212 pages; £30.-

No. 19 ENVIRONMENTAL ASPECTS
OF N-NITROSO COMPOUNDS (1978)
Edited by E.A. Walker, M. Castegnaro,
L. Griciute & R.E. Lyle, 566 pages;
£35.-

No. 20 NASOPHARYNGEAL
CARCINOMA: ETIOLOGY AND
CONTROL (1978)
Edited by G. de-Thé & Y. Ito,
610 pages; £35.-

No. 21 CANCER REGISTRATION
AND ITS TECHNIQUES (1978)
Edited by R. MacLennan, C.S. Muir,
R. Steinitz & A. Winkler, 235 pages;
£11.95

No. 22 ENVIRONMENTAL CARCINO-
GENS - SELECTED METHODS OF
ANALYSIS
Editor-in-Chief H. Egan
Vol. 2 - METHODS FOR THE MEASURE-
MENT OF VINYL CHLORIDE IN
POLY(VINYL CHLORIDE), AIR, WATER
AND FOODSTUFFS (1978)
Edited by D.C.M. Squirrell & W. Thain,
142 pages; £35.-

No. 23 PATHOLOGY OF TUMOURS IN
LABORATORY ANIMALS. VOLUME II.
TUMOURS OF THE MOUSE (1979)
Editor-in-Chief V.S. Turusov, 669 pages;
£35.-

No. 24 ONCOGENESIS AND HERPES-
VIRUSES III (1978)
Edited by G. de-Thé, W. Henle & F. Rapp
Part 1, 580 pages; £20.-
Part 2, 522 pages; £20.-

No. 25 CARCINOGENIC RISKS -
STRATEGIES FOR INTERVENTION
(1979)
Edited by W. Davis & C. Rosenfeld,
283 pages; £20.-

No. 26 DIRECTORY OF ON-GOING
RESEARCH IN CANCER EPI-
DEMIOLOGY 1978 (1978)
Edited by C.S. Muir & G. Wagner,
550 pages; out of print

No. 27 MOLECULAR AND CELLULAR
ASPECTS OF CARCINOGEN
SCREENING TESTS (1980)
Edited by R. Montesano, H. Bartsch &
L. Tomatis, 371 pages; £20.-

No. 28 DIRECTORY OF ON-GOING
RESEARCH IN CANCER EPIDEMIOLOGY
1979 (1979)
Edited by C.S. Muir & G. Wagner,
672 pages; out of print

No. 29 ENVIRONMENTAL CARCINO-
GENS - SELECTED METHODS OF
ANALYSIS
Editor-in-Chief H. Egan
Vol. 3 - ANALYSIS OF POLYCYCLIC
AROMATIC HYDROCARBONS IN
ENVIRONMENTAL SAMPLES (1979)
Edited by M. Castegnaro, P. Bogovski,
H. Kunte & E.A. Walker, 240 pages; £17.50

No. 30 BIOLOGICAL EFFECTS OF
MINERAL FIBRES (1980)
Editor-in-Chief J.C. Wagner
Volume 1, 494 pages; £25.-
Volume 2, 513 pages; £25.-

No. 31 N-NITROSO COMPOUNDS:
ANALYSIS, FORMATION AND
OCCURRENCE (1980)
Edited by E.A. Walker, M. Castegnaro,
L. Griciute & M. Börzsönyi, 841 pages;
£30.-

No. 32 STATISTICAL METHODS IN
CANCER RESEARCH
Vol. 1. THE ANALYSIS OF CASE-
CONTROL STUDIES (1980)
By N.E. Breslow & N.E. Day, 338 pages;
£17.50

No. 33 HANDLING CHEMICAL
CARCINOGENS IN THE LABORATORY
- PROBLEMS OF SAFETY (1979)
Edited by R. Montesano, H. Bartsch,
E. Boyland, G. Della Porta, L. Fishbein,
R.A. Griesemer, A.B. Swan & L. Tomatis,
32 pages £3.95

SCIENTIFIC PUBLICATIONS SERIES

No. 34 PATHOLOGY OF TUMOURS IN LABORATORY ANIMALS. VOLUME III. TUMOURS OF THE HAMSTER (1982)
Editor-in-Chief V.S. Turusov,
461 pages; £30.-

No. 35 DIRECTORY OF ON-GOING RESEARCH IN CANCER EPIDEMIOLOGY 1980 (1980)
Edited by C.S. Muir & G. Wagner,
660 pages; out of print

No. 36 CANCER MORTALITY BY OCCUPATION AND SOCIAL CLASS 1851-1971 (1982)
By W.P.D. Logan, 253 pages £20.-

No. 37 LABORATORY DECONTAMINATION AND DESTRUCTION OF AFLATOXINS B_1, B_2, G_1, G_2 IN LABORATORY WASTES (1980)
Edited by M. Castegnaro, D.C. Hunt, E.B. Sansone, P.L. Schuller, M.G. Siriwardana, G.M. Telling, H.P. Van Egmond & E.A. Walker,
59 pages; £5.95

No. 38 DIRECTORY OF ON-GOING RESEARCH IN CANCER EPIDEMIOLOGY 1981 (1981)
Edited by C.S. Muir & G. Wagner,
696 pages; out of print

No. 39 HOST FACTORS IN HUMAN CARCINOGENESIS (1982)
Edited by H. Bartsch & B. Armstrong
583 pages; £35.-

No. 40 ENVIRONMENTAL CARCINOGENS. SELECTED METHODS OF ANALYSIS
Editor-in-Chief H. Egan
Vol. 4. SOME AROMATIC AMINES AND AZO DYES IN THE GENERAL AND INDUSTRIAL ENVIRONMENT (1981)
Edited by L. Fishbein, M. Castegnaro, I.K. O'Neill & H. Bartsch, 347 pages; £20.-

No. 41 N-NITROSO COMPOUNDS: OCCURRENCE AND BIOLOGICAL EFFECTS (1982)
Edited by H. Bartsch, I.K. O'Neill, M. Castegnaro & M. Okada,
755 pages; £35.-

No. 42 CANCER INCIDENCE IN FIVE CONTINENTS. VOLUME IV (1982)
Edited by J. Waterhouse, C. Muir, K. Shanmugaratnam & J. Powell,
811 pages; £35.-

No. 43 LABORATORY DECONTAMINATION AND DESTRUCTION OF CARCINOGENS IN LABORATORY WASTES: SOME N-NITROSAMINES (1982) Edited by M. Castegnaro, G. Eisenbrand, G. Ellen, L. Keefer, D. Klein, E.B. Sansone, D. Spincer, G. Telling & K. Webb, 73 pages £6.50

No. 44 ENVIRONMENTAL CARCINOGENS. SELECTED METHODS OF ANALYSIS
Editor-in-Chief H. Egan
Vol. 5. SOME MYCOTOXINS (1983)
Edited by L. Stoloff, M. Castegnaro, P. Scott, I.K. O'Neill & H. Bartsch,
455 pages; £20.-

No. 45 ENVIRONMENTAL CARCINOGENS. SELECTED METHODS OF ANALYSIS
Editor-in-Chief H. Egan
Vol. 6: N-NITROSO COMPOUNDS (1983)
Edited by R. Preussmann, I.K. O'Neill, G. Eisenbrand, B. Spiegelhalder & H. Bartsch, 508 pages; £20.-

No. 46 DIRECTORY OF ON-GOING RESEARCH IN CANCER EPIDEMIOLOGY 1982 (1982)
Edited by C.S. Muir & G. Wagner,
722 pages; out of print

No. 47 CANCER INCIDENCE IN SINGAPORE (1982)
Edited by K. Shanmugaratnam, H.P. Lee & N.E. Day, 174 pages; £10.-

SCIENTIFIC PUBLICATIONS SERIES

No. 48 CANCER INCIDENCE IN THE USSR (1983) Second Revised Edition
Edited by N.P. Napalkov, G.F. Tserkovny, V.M. Merabishvili, D.M. Parkin, M. Smans & C.S. Muir, 75 pages; £10.-

No. 49 LABORATORY DECONTAMINATION AND DESTRUCTION OF CARCINOGENS IN LABORATORY WASTES: SOME POLYCYCLIC AROMATIC HYDROCARBONS (1983)
Edited by M. Castegnaro, G. Grimmer, O. Hutzinger, W. Karcher, H. Kunte, M. Lafontaine, E.B. Sansone, G. Telling & S.P. Tucker, 81 pages; £7.95

No. 50 DIRECTORY OF ON-GOING RESEARCH IN CANCER EPIDEMIOLOGY 1983 (1983)
Edited by C.S. Muir & G. Wagner, 740 pages; out of print

No. 51 MODULATORS IN EXPERIMENTAL CARCINOGENESIS (1983)
Edited by R. Montesano & V.S. Turusov, 307 pages; £25.-

No. 52 SECOND CANCER IN RELATION TO RADIATION TREATMENT FOR CERVICAL CANCER: RESULTS OF A CANCER REGISTRY COLLABORATION (1983)
Edited by N.E. Day & J.C. Boice, Jr, 207 pages; £17.50

No. 53 NICKEL IN THE HUMAN ENVIRONMENT (1984)
Editor-in-Chief, F.W. Sunderman, Jr, 529 pages; £30.-

No. 54 LABORATORY DECONTAMINATION AND DESTRUCTION OF CARCINOGENS IN LABORATORY WASTES: SOME HYDRAZINES (1983)
Edited by M. Castegnaro, G. Ellen, M. Lafontaine, H.C. van der Plas, E.B. Sansone & S.P. Tucker, 87 pages; £6.95

No. 55 LABORATORY DECONTAMINATION AND DESTRUCTION OF CARCINOGENS IN LABORATORY WASTES: SOME N-NITROSAMIDES (1983)
Edited by M. Castegnaro, M. Benard, L.W. van Broekhoven, D. Fine, R. Massey, E.B. Sansone, P.L.R. Smith, B. Spiegelhalder, A. Stacchini, G. Telling & J.J. Vallon, 65 pages; £6.95

No. 56 MODELS, MECHANISMS AND ETIOLOGY OF TUMOUR PROMOTION (1984)
Edited by M. Börszönyi, N.E. Day, K. Lapis & H. Yamasaki, 532 pages, £30.-

No. 57 N-NITROSO COMPOUNDS: OCCURRENCE, BIOLOGICAL EFFECTS AND RELEVANCE TO HUMAN CANCER (1984)
Edited by I.K. O'Neill, R.C. von Borstel, C.T. Miller, J. Long & H. Bartsch, 1013 pages, £75.-

No. 58 AGE-RELATED FACTORS IN CARCINOGENESIS (1985)
Edited by A. Likhachev, V. Anisimov & R. Montesano (in press)

No. 59 MONITORING HUMAN EXPOSURE TO CARCINOGENIC AND MUTAGENIC AGENTS (1985)
Edited by A. Berlin, M. Draper, K. Hemminki & H. Vainio 457 pages, £25,-

No. 60 BURKITT'S LYMPHOMA: A HUMAN CANCER MODEL (1985)
Edited by G. Lenoir, G. O'Conor & C.L.M. Olweny (in press)

No. 61 LABORATORY DECONTAMINATION AND DESTRUCTION OF CARCINOGENS IN LABORATORY WASTES: SOME HALOETHERS (1985)
Edited by M. Castegnaro, M. Alvarez, M. Iovu, E.B. Sansone, G.M. Telling & D.T. Williams 55 pages, £5.95

SCIENTIFIC PUBLICATIONS SERIES

No. 62 DIRECTORY OF ON-GOING RESEARCH IN CANCER EPIDEMIOLOGY 1984 (1984)
Edited by C.S. Muir & G.Wagner;
728 pages; £18.-

No. 63 VIRUS-ASSOCIATED CANCERS IN AFRICA (1984)
Edited by A.O. Williams, G.T. O'Conor, G.B. de-Thé & C.A. Johnson, 773 pages, £20.-

No. 64 LABORATORY DECONTAMINATION AND DESTRUCTION OF CARCINOGENS IN LABORATORY WASTES: SOME AROMATIC AMINES AND 4-NITROBIPHENYL (1985)
Edited by M. Castegnaro, J. Barek, J. Dennis, G. Ellen, M. Klibanov, M. Lafontaine, R. Mitchum, P. Van Roosmalen, E.B. Sansone, L.A. Sternson & M. Vahl
85 pages, £5.95

No. 65 INTERPRETATION OF NEGATIVE EPIDEMIOLOGICAL EVIDENCE FOR CARCINOGENICITY
Edited by N.J. Wald & R. Doll
(in press)

NON-SERIAL PUBLICATIONS

(Available from IARC)

ALCOOL ET CANCER (1978)
by A.J. Tuyns (in French only)
42 pages; Fr.fr. 35.-; Sw.fr. 14.-

CANCER MORBIDITY AND CAUSES OF DEATH AMONG DANISH BREWERY WORKERS (1980) By O.M. Jensen
145 pages; US$ 25.00; Sw.fr. 45.-

THE LIBRARY
UNIVERSITY OF CALIFORNIA
San Francisco
(415) 476-2335

THIS BOOK IS DUE ON THE LAST DATE STAMPED BELOW

Books not returned on time are subject to fines according to the Library Lending Code. A renewal may be made on certain materials. For details consult Lending Code.

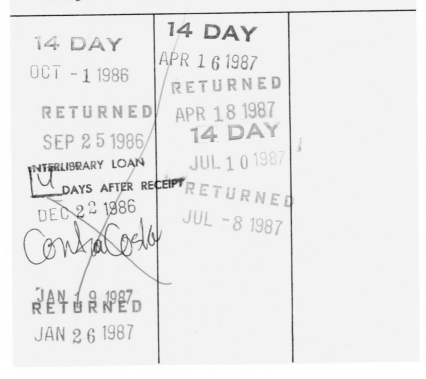